D1037301

FARMERS ON THE ROAD

Interfarm Migration and the Farming of Noncontiguous Lands in Three Midwestern Townships, 1939–1969

Michael D. Sublett

Illinois State University

THE UNIVERSITY OF CHICAGO

DEPARTMENT OF GEOGRAPHY

RESEARCH PAPER NO. 168

1975

Published 1975 by The Department of Geography
The University of Chicago, Chicago, Illinois

Library of Congress Cataloging in Publication Data

Sublett, Michael D.

Farmers on the Road.

(Research paper—University of Chicago, Department of Geography; no. 168)
Bibliography: p. 206

1. Family farms—Middle West. 2. Farmers—Middle West. 3. Farm management—Middle West. I. Title. II. Series: Chicago. University. Dept. of Geography. Research paper; no. 168.

H31. C514 no. 168 [HD1476.U5] 910s [338.1'0977]
ISBN 0-89065-075-6 pbk.
75-8522

Research Papers are available from:
The University of Chicago
Department of Geography
5828 S. University Avenue
Chicago, Illinois 60637
Price: $5.00 list; $4.00 series subscription

ACKNOWLEDGEMENTS

This research effort would have collapsed without the willing cooperation of more than 700 informants. While they must remain anonymous, I will never forget what they did for me. A few men, who in their official capacities were especially helpful during the course of the field work, can be mentioned. They include: Vernon Evans, Real Estate Specialist for the U.S. Army Corps of Engineers; Burch Harrington, Balanced Farming Agent in Putnam County; Harold Holz, Land Manager at the Joliet Arsenal; William Polley, ASCS Office Manager in Joliet; and William Richardson, ASCS Office Manager in Unionville. A number of farm equipment manufacturers generously provided me with photographs of their products.

From the outset, through all the ups and downs, my research has been ably, patiently, and kindly overseen by Wesley C. Calef. William D. Pattison, my second reader, helped immeasurably to improve the final version of the manuscript. Paul F. Mattingly and several of my other colleagues here at Illinois State University have in one way or another been of assistance as I was writing up my findings. To all of these professional geographers, I am grateful. Finally, I offer my thanks, although inadequate, to my wife, Patty, for her countless contributions.

TABLE OF CONTENTS

LIST OF TABLES

LIST OF FIGURES

LIST OF PLATES

CHAPTER I

INTRODUCTION

The general subject of this study is one of the most efficient American units of production, the family farm.[1] Although corporate competition is intense in certain types of agriculture, the bulk of the nation's raw food and fiber originates on farms operated by a single family. As an institution the family farm is therefore fairly secure, but the agricultural survival of any given farmer might be doubtful. Not infrequently in the last decade or so the inability of many good farmers to remain on the land has been due in large measure to the infamous cost-price squeeze. Sophisticated explanations of the squeeze are obtainable, most of whose authors agree the farmer is simply caught (seemingly through no fault of his own) between the relatively low prices being offered by society for his products and the high prices charged him for his inputs, such as fertilizer, machinery, and land. According to one knowledgeable observer "the 'terms of trade' between agriculture and the remainder of the national economy have turned against agriculture; twice as much output is now required to provide the same relative income as was true 50 years ago."[2]

Some Responses to the Squeeze

Several alternatives, short of capitulation, are available to individuals seeking to avoid the ravages of the squeeze. Two of these alternatives, moving to another farm and farm enlargement, plus a closely related action, the farming of noncontiguous land, are the subjects of detailed analysis at the local level in this study. The general setting of the study is the American Midwest during the period, 1939-69. In Chapter II we will identify and describe the areas chosen for scrutiny. Chapter III explains the various information sources that were employed. Interfarm migration is the subject of Chapter IV, while the last three

---

[1]On a family farm, most of the labor and all the management are provided by the farmer, his wife, and their children.

[2]Marion Clawson, Policy Directions for U.S. Agriculture: Long-Range Choices in Farming and Rural Living (Baltimore: Resources for the Future, 1968), pp. 189-90.

substantive chapters (V, VI, and VII) are devoted to a discussion of farm expansion through acquisition of nonfarmstead land and the operation of such farms. Chapters VI and VII deal specifically with the difficulties encountered by farmers having fragmented farm units. Meanwhile, in the balance of this introductory chapter we will be exploring (1) several of the typical responses to the cost-price squeeze including the two mentioned above that will be of special moment to us later, (2) literature pertaining to interfarm migration and noncontiguous farming, (3) the evolution of this author's interest in the topics to be considered, and (4) some of the main questions to be addressed in this study.

Doing Custom Work

A typical midwestern farmer, although annually underemployed, must overmechanize to compensate for severe seasonal demands on his time. Consequently, such an overmechanized farmer may choose to hire out his machine and himself to perform field chores like plowing, haying, and combining for other farmers so as to exploit more fully both his and the machinery's potential. Custom work, it should be noted, requires long hours in the field when conditions are just right. So unless he plans carefully, a custom operator can find himself neglecting his own farm to keep good customers satisfied.

Working Off the Farm

Rejecting the custom-hire alternative, the farmer might take a job off the farm.[1] If the job is full time, he must schedule field tasks, equipment maintenance, livestock chores, and maybe a short course on new fertilizer techniques around a fairly rigid work timetable. Now instead of too much time, he has too little. If off-farm employment is to occupy him only in the off-season, he must usually accept a temporary status and thereby forfeit the advantages bestowed by seniority. When the wife is willing, some families find it preferable to have her work away, expecting the farmer to assist around the house. But even if the farmer is doing something profitable besides farming, that idle equipment is not helping pay for itself.

---

[1]"Since 1970 the [American] farm population has received more of its net income from off-farm than from home-farm sources." U.S., Department of Agriculture, Economic Research Service, The One-Man Farm, by Warren R. Bailey, ERS-519 (Washington: Government Printing Office, 1973), p. 5.

Contracting

A growing proportion of farmers bargains away part of its managerial prerogatives by contracting with a processing firm to supply it with some agricultural commodity. One can agree to plant a specified crop and deliver the harvest to the processor in return for a guaranteed price. One can have a 15,000-bird hen house left in the front yard with the understanding that the contracting firm will expect a certain quantity of eggs per day. The processing firm gets a dependable source of raw materials and the farmer a dependable supplementary income source. Farmers with large families can keep the younger kids busy in the hen house earning their keep without exposing them too soon to the dangers of working with heavy machinery. A few integration schemes, like popcorn contracting, also allow the farmer to utilize more effectively his present inventory of equipment. The desire of a man to contract will be unfulfilled, of course, if he should happen to live where there are no firms seeking partners.

So far we have restricted the farmer to the land he is already farming. What new possibilities for adjustment appear if we lift the land restriction? First, the man can now shift from farm to farm. Second, he can expand his acreage while remaining headquartered at his established base. The following paragraphs discuss these two alternatives.

Moving to Another Farm

If he chooses the migration alternative, the farmer might be seeking better land, more land, a different landlord-tenant arrangement, or any number of other advantages. Maybe the old landlord is impossible to please, or the time has finally come to buy a small place for a permanent farmstead. Perhaps the new farm offers a cattle-feeding setup or more tillable acres than the farmer had before. In some cases, as we shall see later, relocation of midwestern farm families has been precipitated by reasons other than the desire for survival or betterment; but a majority of interfarm movements occur because farmers expect to improve themselves on another place.

Farm Enlargement

Acreage expansion from an established base is for many the most appealing and realistic response to the cost-price squeeze. Repeated studies of farmers' records tell us that it pays the farmer with a fixed line of equipment and amount of labor to push up his acreage or volume of production to the point where average cost per unit reaches a minimum.[1] Fully mechanized one-man and two-man farms can pretty well exhaust all the size economies with the proper mixture of inputs so that there is no pressing reason from a cost-minimization standpoint to expand beyond this stage. On larger farms that do surpass the minimum cost size the farmers may be seeking instead to maximize profit. They must endure more risks; but because of their size they can take advantage of discounts available on bulk purchases, better marketing arrangements, more financial leverage in the credit marketplace, and tax loopholes.[2]

The expanding farmer has for some time been recognized as the greatest single factor in the land purchase market. Purchases for farm enlargement in the conterminous United States have increased from 26 percent of all transfers (of land to be farmed) in the period, 1950-54, to 62 percent in the year ending March 1, 1971.[3] In the Corn Belt states, the expander has been even more aggressive than his national counterpart, accounting in the above periods for 28 percent and 68 percent of transfers, respectively. Heretofore, the expander's role in the land rental market has been less well-known because of that market's personal, extralegal nature. One of our aims here is to make it better known. Since the tenure status of farm operators will be of concern throughout this report, a word about the commonly recognized tenure classes is appropriate. Some farmers own all the land they farm. We will refer to them as owner-operators. Others rent all their land from one or more landlords. These farmers are generally known as tenants. Still others, usually called partowners, rent some of their land and own the rest.

As a farmer's acreage increases so does the likelihood of his having to contend with noncontiguous tracts. Simple is the fact of farm life in the Midwest that land adjacent to the farmstead tract does not always become available just when a farmer wants nor is he always ready to acquire land if it should

---

[1]For a clear explanation of scale economies, a summary of several studies, and a lengthy bibliography see: U.S., Department of Agriculture, Economic Research Service, Economies of Size in Farming, by J. P. Madden, Agricultural Economic Report 107 (Washington: Government Printing Office, 1967).

[2]U.S., Department of Agriculture, Economic Research Service, Midwestern Corn Farms: Economic Status and the Potential for Large and Family-Sized Units, by Kenneth R. Krause and Leonard R. Kyle, Agricultural Economic Report 216 (Washington: Government Printing Office, 1970).

[3]Data on transfers are from: U.S., Department of Agriculture, Economic Research Service, Farm Real Estate Market Developments (Washington: Government Printing Office, March, 1969), p. 14; (August, 1971), p. 36.

come up for rent or sale. In the portion of this inquiry devoted to farm enlargement, our primary concern will be with the patterns and problems of noncontiguity. After all, few expanding farmers have been able to avoid it.

Reviewing the Literature: Interfarm Migration

American farmers moved regularly in the nineteenth century; but some were nothing more than petty speculators who scratched out a living on a few acres until, according to frontier historian Le Duc, they "sold out and moved on to repeat the speculative cycle. . . ."[1] Gates, another frontier historian, felt sheer restlessness probably accounted for a good deal of the turnover. "Settlers hearing of better opportunities elsewhere, perhaps in Minnesota or in Dakota Territory, might sell a relinquishment [to their Iowa homestead], receiving the value of their improvements and of their claim and move on."[2] Others, who were casualties of grasshopper plagues, dry spells, or lack of credit during times of low prices, would relocate and hope for better luck next time.[3] Some contemporaries, it seems, took a dim view of the transient nature of their

---

[1]Thomas Le Duc, "Public Policy, Private Investment, and Land Use in American Agriculture, 1825-1875," Agricultural History 37 (January, 1963): 6.

[2]Paul W. Gates, "The Homestead Law in Iowa," Agricultural History 38 (April, 1964): 77.

[3]Well-known to many American children are Laura Ingalls Wilder's delightful accounts of the migrations she, her parents, and her sisters made in the 1870's searching for the right place to farm and of the events that transpired between moves. The saga of the Charles Ingalls family began in western Wisconsin with The Little House in the Big Woods which Laura published in 1932. From there he took the family south to a homesite on the banks of the Verdigris River in what is now Oklahoma but was then Indian Territory. This first migration and the year that followed were chronicled by Laura in the most widely known of the series, Little House on the Prairie (1935). When the federal government forbade squatting in that area, the family headed northward to settle in western Minnesota. This period was described by Laura in On the Banks of Plum Creek (1937). Late in the 1870's after grasshoppers and poor yields had taken their toll, Ma and Pa Ingalls decided to follow the railroad west to a homestead near De Smet, Dakota Territory. In and near De Smet the last four books in the original series are set. They include: By the Shores of Silver Lake (1939); The Long Winter (1940); Little Town on the Prairie (1941); and These Happy Golden Years (1943). Following Laura's death in 1957 at the age of 90, two additional works appeared. The First Four Years (1971) describes her early married life near De Smet; On the Way Home (1962) is based on her diary of the 1894 wagon trek by Laura, her husband (Almanzo Wilder), and their daughter to the place in the Missouri Ozarks where they spent the balance of their lives together. Though intended for children, the "Little House" series should not be overlooked by a serious student of the American frontier.

neighbors. The following query appeared in an 1858 issue of Moore's Rural New Yorker:

> Why ditch, fence, put out fruit trees, and put up good buildings on the homestead, when manifest destiny has arranged for your departure to California or the moon, next year or the year after?[1]

A truly thorough examiner of post-Civil War farmer turnover was the Great Plains historian Malin. Working in selected Kansas townships with manuscript lists from federal censuses to 1885 and state census returns for later years, he was able to determine the length of time a farmer remained in the township and when he left or retired. Turnover was large during the pioneering period, but by 1915 farm population had reached "a high degree of stabilization."[2] Since mobility was "the most conspicuous (tenure) characteristic" in the nineteenth century, land legislation such as the Homestead Act of 1862 requiring actual occupance and construction of a dwelling had to contend "against overwhelming public opinion."[3] "The difficulty under the Homestead law was that it violated the mobility factor in prescribing the five-year residence."[4] It would have worked badly, he believed, on any frontier and not just along the semi-arid one encountered in Kansas.

Malin carried his study of turnover through the 1930's and formed an opinion about the Great Depression which differed from that held by many social scientists of the era. In Kansas, and perhaps by extension elsewhere, farm population tended, he felt, toward locational stability (within a township) during periods of economic depression and toward mobility at times of prosperity.[5] During a depression, he explained, "there is little opportunity to sell to advantage, or even for a mortgagee to force a sale, but with rising prices the owner is glad to take his profits, if any, and move on."[6] Furthermore, "the propaganda to the contrary notwithstanding," the 1930's seemed to have no greater

---

[1] Moore's Rural New Yorker (March 13, 1858), p. 85. Cited by Kathleen A. Smith, "Moore's Rural New Yorker: A Farm Program for the 1850s," Agricultural History 45 (January, 1971): 41.

[2] James C. Malin, The Grassland of North America: Prolegomena to its History (Lawrence, Kansas: By the Author, 1947), p. 281.

[3] James C. Malin, "Mobility and History: Reflections on the Agricultural Policies of the United States in Relation to a Mechanized World," Agricultural History 17 (October, 1943): 180.

[4] Ibid., p. 181.

[5] Malin, Grassland, p. 282.

[6] Malin, "Mobility and History," p. 182.

effect on the mobility-stability ratio than earlier periods of hard times. In fact, there were probably "fewer farmers on the move, proportionally, than in any previous depression in Kansas."[1] It is unfortunate from our standpoint that he did not have access to data on actual farm-to-farm movement within each of his census tracts.

Whether the Great Depression produced an upsurge of farmer turnover or not, the possibility provoked much professional interest. Wakeley, for instance, compared the migration habits of open-country Iowans with those of village dwellers for the period, 1928-1935.[2] At the same time, in a virtually identical project, Lively was studying rural Ohio migration patterns.[3] Both men employed data gathered in selected townships by enumerators working for the federal Emergency Relief Administration (ERA). The period chosen for investigation "was short enough," wrote Wakeley, "to encourage accuracy and long enough to indicate changes made during the depression."[4] Wakeley, whose report is much more comprehensive than that of his Ohio counterpart, searched the ERA data for answers to a variety of questions. How often did families move? What proportion of the total population were the movers? How far had they moved? At what age did the older children leave home and where did they go? Was there a relationship between tenure status and migration? Were families on relief as mobile as other families?[5] There was a tendency, he discovered, "for farmers to move more frequently than villagers" and for tenants to do "most of the moving."[6] Owner-operators, on the other hand, were inclined to migrate longer distances than those who did not own the land they farmed.

After mentioning several previous mobility studies that he felt ignored local movements, Lively discussed the 1928-1935 circulation patterns of persons living in the sample of ten Ohio townships as of January 1, 1935. He, too, plotted distance and frequency of moves made by entire families and the dispersion habits of children leaving home for the first time. He did not, however,

---

[1]Malin, Grassland, p. 282.

[2]Ray E. Wakeley, Differential Mobility Within the Rural Population in 18 Iowa Townships, 1928 to 1935, Iowa Agricultural Experiment Station Research Bulletin 249 (December, 1938), pp. 287-318.

[3]C. E. Lively, "Spatial Mobility of the Rural Population with Respect to Local Areas," American Journal of Sociology 63 (July, 1937): 89-102.

[4]Wakeley, Differential Mobility, p. 280.

[5]Ibid., p. 279.

[6]Ibid., pp. 292-93.

separate farmers from the townships' residents who did not farm. Short distance moves, as expected, were the rule both for conjugal families and their grown offspring.[1] State boundaries, as it turned out, had a noticeable effect on circulation. In cases where the sample township was located in a county next to a neighboring state "it was evident that the state line served as a sort of barrier to rural migrants."[2] Like Malin and Wakeley, Lively was handicapped in his research effort by lack of data on intratownship moves. How, one wonders, was it possible to study "the circulation of population with respect to local areas"[3] without considering at all those moves which took place within the bounds of the ten townships?

Illinois farmers were also on the move. To learn why, Case and Warren obtained statements from 536 northern Illinois farm tenants who changed residence on or about March 1, 1941.[4] Over half said they moved because the owner or a relative of the owner wanted the farm or because they believed they could better themselves by relocating. The authors considered this a healthy sign. Perhaps there was, after all, less of an unwelcome effect on "the economic well-being of the [mobile] family" and less disruption of community life than previously suspected.[5] A few years later Salter reported another team of investigators had come to essentially the same conclusion in their study of Clay County, Nebraska.[6]

Among those subscribing to the notion that tenants suffered from excessive moving was Cotton. Tenure instability, he suggested in a 1937 article, had led to resource waste. "There are serious losses involved in the frequent moving of tenants" plus certain "undesirable results in rural social life."[7] Also

---

[1] Lively, "Spatial Mobility," p. 89.

[2] *Ibid.*, p. 100.

[3] *Ibid.*, p. 89.

[4] H. C. M. Case and S. I. Warren, "Why Farm Tenants Move," *Illinois Farm Economics*, No. 88 (September, 1942), pp. 366-70.

[5] *Ibid.*, p. 366.

[6] L. F. Garey, G. H. Lambrecht, and Frank Miller, *Farm Tenancy in Clay County, Nebraska*, Nebraska Agricultural Experiment Station Bulletin 337 (1942), mentioned by Leonard A. Salter, *A Critical Review of Research in Land Economics* (Madison: University of Wisconsin Press, 1967), p. 185. Salter's study first appeared in 1948.

[7] Albert H. Cotton, "Regulations of Farm Landlord-Tenant Relationships," *Law and Contemporary Problems* 10 (October, 1937): 509.

concerned about rural upheaval in the late 1930's were several community specialists operating under the aegis of the Bureau of Agricultural Economics, United States Department of Agriculture. They scattered, in 1939, to six areas of the country to act as participant observers and later reported their findings in a series of monographs known collectively as the Rural Life Studies.[1] As expected, they found the greatest degree of tenure stability among the Amish of eastern Pennsylvania. Least stable were farmers in the area around Sublette, Kansas.[2]

Duncan looked at the causes, kinds, volume, and implications of migratory movements by southern (primarily Oklahoman) farm people.[3] Moves were classified according to the distance traversed and the proportion of community ties that had to be broken as a result. Data on short-distance moves proved difficult to obtain, but as far as he could tell these movers failed "to obey any law of direction."[4] Apparent, however, was a strong attachment for particular localities. Families had a tendency to move no farther than necessary so as to avoid severance of links with friends, relatives, local institutions, and the type of land and farming they knew best. Longer moves, whether to another county,

---

[1]Glen Leonard and C. P. Loomis, Culture of a Contemporary Rural Community: El Cerrito, New Mexico, Rural Life Study No. 1 (Washington: U.S. Department of Agriculture, Bureau of Agricultural Economics, 1941); Earl H. Bell, Culture of a Contemporary Rural Community: Sublette, Kansas, Rural Life Study No. 2 (Washington: U.S. Department of Agriculture, Bureau of Agricultural Economics, 1942); Kenneth MacLeish and Kimball Young, Culture of a Contemporary Rural Community: Landaff, New Hampshire, Rural Life Study No. 3 (Washington: U.S. Department of Agriculture, Bureau of Agricultural Economics, 1942); Walter M. Kollmorgen, Culture of a Contemporary Rural Community: The Old Order Amish of Lancaster County, Pennsylvania, Rural Life Study No. 4 (Washington: U.S. Department of Agriculture, Bureau of Agricultural Economics, 1942); Edward O. Moe and Carl C. Taylor, Culture of a Contemporary Rural Community: Irwin, Iowa, Rural Life Study No. 5 (Washington: U.S. Department of Agriculture, Bureau of Agricultural Economics, 1942); Waller Wynne, Culture of a Contemporary Rural Community: Harmony, Georgia, Rural Life Study No. 6 (Washington: U.S. Department of Agriculture, Bureau of Agricultural Economics, 1943).

[2]The Sublette area has been studied again by William E. Mays, Sublette Revisited: Stability and Change in a Rural Kansas Community after a Quarter of a Century (New York: Florham Park Press, 1968).

[3]Otis Durant Duncan, "The Theory and Consequences of Mobility of Farm Population," in Population Theory and Policy, ed. by Joseph J. Spengler and Otis Dudley Duncan (Glencoe, Illinois: The Free Press, 1956). Duncan's paper originally appeared as: Oklahoma Agricultural Experiment Station Circular 88 (May, 1940).

[4]Duncan, "The Theory," p. 420.

to a nearby city, or all the way to California "tended to be destructive of existing social bonds and established relationships."[1]

McKain is the coauthor of two farmer turnover studies. With Metzler he published an article based on a survey of wartime Agricultural Adjustment Administration records kept on farmers in Solano County, California.[2] Of the farmers who left 1943 farms, over three-quarters were by 1944 operating other farms in the same county. The authors concluded theirs was an efficient method of tracing turnover if done systematically on an annual basis. Furthermore, the data obtained on turnover and retirement would aid "post-war planning committees who [might] need this sort of information."[3] Another Bureau of Agricultural Economics assignment led McKain and Dahlke to the Snake River Valley for a detailed review of farmer turnover in an irrigation project.[4] Between the date of initial settlement after delivery of the water (sometime in the 1930's) and 1945, the annual rate of operator turnover averaged about 10 percent with slightly lower rates prevailing before 1941 and higher afterward. Many of the operators by 1945 were tenants. This the authors felt contributed to turnover as did the emphasis on specialty crops which made project farmers susceptible to critical fluctuations in farm income from year to year.[5]

Bremer, a historian, has just published a thoughtful account of farmer mobility for six central Nebraska townships.[6] Using mainly personal property tax records, he has calculated what he calls "gross persistence rates" of township farmers for periods ranging from five to forty years. In some ways this study is an updating of those done by Malin several decades earlier. Ironically, Bremer too is unable to solve the problem of intratownship movement. On this subject he writes:

---

[1] Ibid., p. 433.

[2] Walter C. McKain, Jr. and William H. Metzler, "Measurement of Turnover and Retirement of Farm Owners and Operators," Rural Sociology 10 (March, 1945): 73-76.

[3] Ibid., p. 76.

[4] U.S., Department of Agriculture, Bureau of Agricultural Economics, Turn-over of Farm Owners and Operators, Vale and Owyhee Irrigation Projects, by Walter C. McKain, Jr. and H. Otto Dahlke (Berkeley, California: n.p., 1946).

[5] Ibid., p. 13.

[6] Richard G. Bremer, "Patterns of Spatial Mobility: A Case Study of Nebraska Farmers, 1890-1970," Agricultural History 48 (October, 1974): 529-42.

> some farm operators moved from one farmstead to another within the same township. Studies based on census records would miss this type of movement and thus understate the degree of turnover. Local records proved inadequate to remedy this problem. In some instances personal property tax records bore the wrong section location numbers while in other cases no such locations appeared at all. Since many farm operators rented their land, deed and mortgage records offered no solution to this problem. Consequently if the farmer remained in his township he was counted as persistent no matter where his listed residence.[1]

Despite this and other difficulties, Bremer's is a contribution to our understanding of twentieth-century farmer turnover patterns in the Great Plains. Particularly noteworthy is his attempt to explain why the persistence of farmers has been increasing during the 1950's and 1960's.[2]

One outgrowth of concern in America over excessive farmer turnover has been an eagerness to recognize families that have retained continuous title to tracts of land for lengthy periods of time. Unofficial recognition might take the form of a complimentary magazine or newspaper article.[3] Officially, several states have sponsored programs to identify and honor farm tracts that have been in the family more than a century. Wisconsin initiated such a program in preparation for its 1948 centennial celebration. Tarver used questionnaire responses from 774 centennial families and interviews with fifty-eight of them to write his dissertation on farm succession.[4] By the late forties, Indiana's State Historical Society had begun collecting names of Hoosier families who had owned their land more than 100 years. As of September, 1947, over 700 had been found with Wayne County's Quakers providing the greatest concentration.[5] Illinois had a Centennial Farm Program for only a few months in 1972, but the state's Department of Agriculture succeeded in enrolling more than four thousand tracts before the defeat of incumbent Governor Richard Ogilvie in November, 1972, brought the project to an abrupt halt. In Illinois as elsewhere, the intention was to recognize continuous ownership; but, in fact, many tracts were discovered where operatorship by members of the family had also been uninter-

---

[1] Ibid., p. 533.

[2] Ibid., pp. 536-41.

[3] See for instance: "Two Centuries on this Farm," Hoard's Dairyman, June 10, 1942, p. 312.

[4] James Donald Tarver, "Wisconsin Century Farm Families: A Study of Farm Succession" (unpublished Ph.D. dissertation, University of Wisconsin, 1950).

[5] "There's No Wanderlust Among Hoosier Farmers," Logansport Pharos-Tribune and Press, September 30, 1947, p. 3. Cited hereafter as Pharos-Tribune.

rupted. Such a case is described elsewhere by this author.[1]

There have been three recent monographs dealing in some way with turnover in Great Britain. Williams chose a harsh agricultural environment in southwestern England for the locale of his study.[2] Initially, he had aimed "to examine the effects of rural depopulation as a process on the structure of family and kinship" in a stable community.[3] He soon found, however, that although rural life there seemed to be in a state of equilibrium, it actually could be better described as "dynamic equilibrium."[4] And one of the major forces at work perpetuating this condition was the constant shifting around of farmers as they entered the parish, left, or moved from one parish farm to another. By means of interviews and careful perusal of public and church records, he attempted to piece together the landholding pattern back as far as 1900. Turnover was slight up through World War II; but in the period, 1945-55, newcomers ("Up Country Johnnies" was the local appellation) from areas to the east with little or no farm experience began buying the poorest farms in the parish. Many of them quickly went broke, causing more turnover. After 1955, outsiders stopped arriving and those who survived the first few years settled down.[5] A number of informative maps enhance the text of the Ashworthy study. Of special interest here is one depicting holdings by number of occupiers from 1900 to 1960 and its companion on the facing page entitled "Length of Occupation of Holdings, Ashworthy 1960."[6]

The agricultural economist Gasson attempted in the mid-1960's "to measure the increasing and changing impact of London on the types of persons owning and occupying farms in parts of South East England."[7] Available to her were detailed records from the National Farm Survey of 1941 and returns from a mail questionnaire sent out by her college to the occupier of each holding in

---

[1] Michael D. Sublett, William D. Walters, and Southard M. Modry, Commentary on a Corn Belt Countryside: A Self-Guided Rural Experience (Normal, Illinois: Department of Geography-Geology, Illinois State University, 1973), pp. 77-79.

[2] W. M. Williams, A West Country Village, Ashworthy: Family, Kinship and Land (London: Routledge & Kegan Paul, 1963).

[3] Ibid., p. xiv.

[4] Ibid., p. xviii.

[5] Ibid., p. 33.

[6] Ibid., pp. 34-35.

[7] Ruth Gasson, The Influence of Urbanization on Farm Ownership and Practice, Studies in Rural Land Use, Report No. 7 (Ashford, Kent: Department of Agricultural Economics, Wye College, 1966), p. iii.

the sample parishes.[1] Were Londoners flocking out to acquire farms where they could enjoy life alongside a major rail line leading back to the city? The length-of-tenure data seemed to suggest an affirmative answer. Tenants, according to the 1964-65 survey but contrary to the 1941 findings, had been on their farms longer than neighboring owner-operators; and full-time farmers as a group had been there longer than part-timers.[2] A substantial number of the current owner-operators farmed only part of the time because they were really "business and professional men who [did] . . . not need to rely on farming income for their living."[3]

The third of these recent British works was authored by Nalson, a sociologist.[4] Like Williams, he chose a problem area. Although he never identifies it, the bits of information provided point toward the western slopes of the Pennines.[5] Important topics like length of tenure and origin of newcomers are covered quite thoroughly. Most intriguing to this author is his analysis of the effect that altitude has had on the tenure picture. Whether a man was moving uphill or downhill revealed much, according to Nalson, about his community status.

> We know that wide changes in type, quality and size of farms can occur over short distances due to the marked influence of altitude on the farming in the area. A measure of the movement of farmers uphill or downhill compared to where their parents lived will give therefore some indication of changes in farming status which have occurred.[6]

He found, for example, that nearly all farmers (and their wives) whose parents lived outside the survey area had moved uphill (to poorer land) compared to the altitude where their parents lived. These young farmers would start at the top and perhaps work their way down to better land if they decided to stay in the area.[7]

In summary, migration researchers in the United States have tended to concentrate either on the frontier and post-frontier turnover activity or on the apparent upsurge caused by the depression and war that occurred in the fourth

---

[1] *Ibid.*, pp. 8. and 80-81.

[2] *Ibid.*, Chaps. II and III.

[3] *Ibid.*, p. 32.

[4] J. S. Nalson, *Mobility of Farm Families: A Study of Occupational and Residential Mobility in an Upland Area of England* (Manchester: Manchester University Press, 1968).

[5] *Ibid.*, Chap. II.

[6] *Ibid.*, p. 73.

[7] *Ibid.*, pp. 74-77.

and fifth decades of this century. Several had access, unfortunately, only to data at the township level and thus could do nothing with intratownship movers. Since the immediate postwar years, too little effort has been expended in this country on the subject of farmer mobility. Such has not been the case, however, in Great Britain where a trio of in-depth investigations have been conducted recently. Of the British authors, Williams in particular should be lauded for his attention to the spatial side of rural affairs.

Reviewing the Literature: Farming of Noncontiguous Land

That fragmentation of farm operating units persists in Europe and elsewhere around the world is widely known. For those who are interested in the European problem and attempts to remedy it, there exists a surfeit of solid scholarly studies upon which to draw. One might begin with Gregor, Dovring, or Chisholm for the overview and then consult Erhart or Lambert, for instance, to learn in detail about specific examples.[1] The United States has been spared the severe sort of fragmentation that has plagued many other agricultural societies because a major culprit, partition of land among heirs, is avoided. If possible, when there have been multiple heirs to a tract of midwestern farmland and no will to provide guidance, one heir has been permitted by the courts and encouraged by the community to take over the land and to reimburse the other heirs with cash. Should the heirs fail to agree on a suitable solution, the probate court has the authority to order a public auction of the property to clear the estate and keep the tract intact. Despite these strong safeguards against fragmentation and despite what Grigg says,[2] physical separation of farmstead tracts from affiliated nonfarmstead tracts has been, is, and will continue to be a part of American agriculture.

---

[1]Howard F. Gregor, Geography of Agriculture: Themes in Research (Englewood Cliffs, N.J.: Prentice-Hall, 1970), pp. 78, 90, 104, 144-45; Folke Dovring, Land and Labor in Europe in the Twentieth Century (The Hague: Martinus Nijhoff, 1965), pp. 39-56; Michael Chisholm, Rural Settlement and Land Use (New York: John Wiley & Sons, Inc., 1967), pp. 13, 50-51, 69, 71, 131; Rainer R. Erhart, "Land Consolidation and its Effects on the Agricultural Landscape of the Vogelsberg, Germany" (unpublished Ph.D. dissertation, University of Illinois, 1967); Audrey M. Lambert, "Farm Consolidation in Western Europe," Geography 48 (January, 1963): 31-50.

[2]"In England or the United States the typical farm consists of one block of land. . . . " David Grigg, The Harsh Lands: A Study in Agricultural Development (London: Macmillan, 1970), p. 139.

Outlying tracts were on the scene in rural America long before 1939. Of colonial plantations in Maryland and Virginia, Brown wrote, "many planters owned or rented scattered parcels of land. . . ."[1] In Florida, Anderson found evidence that big antebellum plantations often had numerous scattered tracts, sometimes miles from the headquarters tract.[2] Of the distant woodlot used by the midwestern prairie farmer in the nineteenth century, Barrows said, "it was found easier to haul fuel and rails [for fencing] a few miles than to clear forest land, and prairie farmers often purchased a small piece of land for timber in the nearest woods."[3] In his analysis of woodland and prairie in an Iowa county, Hewes noted that prairie settlers continued to depend on riverine timber even after coal became available. His informants, in 1950, could recall their fathers hauling freshly cut firewood several miles and even buying new timber lots as late as 1900.[4]

Scant scholarly attention was afforded the distribution of tillable and grazing tracts within the midwestern farm operating unit prior to about 1940. At that time a flurry of interest in the noncontiguous tract developed following the rapid adoption of pneumatic rubber tires for farm tractors during the 1930's. Looking back at this flurry, land economist Salter viewed it as a modest initial attempt to learn more about a situation that he felt might burgeon into a matter of major importance to farmers and their consultants.[5] Salter was a bit premature with his speculation, because even though the practice of farming away from home waxed, interest in it waned.

Among several items cited by Salter was a Purdue University publication analyzing landholding and land use patterns in two Indiana areas.[6] The

---

[1]Ralph H. Brown, Historical Geography of the United States (New York: Harcourt, Brace & World, 1948), p. 59.

[2]James R. Anderson, "The Beginnings of Plantation Agriculture in North Central Florida," paper read at meeting of the Southeastern Division, Association of American Geographers, November, 1971.

[3]Harlan H. Barrows, Geography of the Middle Illinois Valley, Illinois State Geological Society Bulletin 15 (1910), p. 80.

[4]Leslie Hewes, "Some Features of Early Woodland and Prairie Settlement in a Central Iowa County," Annals of the Association of American Geographers 40 (March, 1950): 51-54.

[5]Salter, A Critical Review of Research in Land Economics, p. 172.

[6]John R. Hays, Relationship of Character of Farming Units to Land Management in Two Townships in Indiana, Purdue Agricultural Experiment Station Bulletin 450 (August, 1940), pp. 13-20.

author Hays examined noncontiguous tracts from several angles, but a primary aim was to reveal what he expected to be a causal relationship between distance from farmstead to outlier on one hand and land management practices on the other. Interest in tract-to-tract connectivity was the theme of a 1942 Ohio State report authored by Headington and Falconer.[1] They placed special emphasis, however, on the reasons farmers gave for wanting more land rather than on the role of distance in the decisions to acquire or not to acquire parcels. Diller's main concern in his study of the Diller home area was with farm tenancy and the problems caused the farmer by the survey system. He did, nevertheless, broach the subject of scattered fields and sketched for us a map of holding patterns at the end of his nineteenth appendix.[2] The spatial intricacies of land tenure patterns even found their way to the pages of Harpers' Magazine in the early 1940's as Taylor mourned the passing of the homestead farm.[3] He blamed the apparent collapse of the agricultural ladder on the ability of greedy men to farm more land and to whiz across the countryside on their pneumatic tires in search of it.

After the abortive pre-World War II fascination with noncontiguity, little more is heard until the mid-1950's when two groups, operating independently, revived the subject. One group, led by the geographer Prunty at the University of Georgia, has focused its attention on the multiple-unit plantation. Members of the Georgia group have monitored the transformation of the plantation from an ownership unit broken into numerous parts and operated by different tenants and croppers to a management unit, still broken into separate tracts, but now operated by a single farmer and his hired help. The other group has been keeping us posted on the disintegration of the midwestern farmscape into an even more complex tangle of tenure ties than those about which Hays, Headington, Falconer, Diller, and Taylor wrote.

In a 1956 paper, Prunty called on the Census Bureau to revise its definition of the multiple-unit farm (southern plantation) and to take note, among other things, of the fact that, "there are certain types of multiple-unit operations in

---

[1] R. C. Headington and J. I. Falconer, Size of Farm Units as Affected by the Farming of Additional Land, Ohio Agricultural Experiment Station Bulletin 637 (October, 1942).

[2] Robert Diller, Farm Ownership, Tenancy, and Land Use in a Nebraska Community (Chicago: The University of Chicago Press, 1941), p. 178.

[3] Paul Schuster Taylor, "Good-By to the Homestead Farm: The Machines Advance in the Corn Belt," Harpers' Magazine, May, 1941, pp. 589-97.

which the units under one management are not contiguous. . . ."[1] Since 1956, a number of Georgia geographers working with southern agricultural occupance have commented on the multiple-unit phenomenon. Only two, Fisher and Aiken, will be mentioned here. While developing a rural occupance classification for the central Georgia Piedmont in the mid-1960's Fisher wrote:

> Rather than a multiple-unit operation being formed of one landholding, that of today is composed of several landholdings. This is a form of functional fragmentation which reaches beyond the limits of the owned landholding. In a sense, it is a functional aggregation of areally disparate production units.[2]

Later, Fisher took the opportunity to look even more closely at a major factor in the functional fragmentation of the South, namely the "widespread manipulation of allotments" assigned by the Agricultural Stabilization and Conservation Service to land that had a history of having recently produced cotton, tobacco, or peanuts.[3] As expected, he found allotments were either being rented separately or entire tracts were being leased by eager farmers just to obtain those few precious allotment acres.

Aiken picks up where Fisher stopped and delivers a polished but preliminary report on the southern version of the large, disconnected farm operating unit.[4] He reviews for his readers the internal fragmentation that characterized plantations during the tenant-cropper era and acknowledges the identification by Prunty of the neoplantation under single ownership before launching his own discussion of the even newer "fragmented neoplantation." After teasing us with just two examples of this newest occupance type, he closes with a series of intriguing but unanswered questions. How, for instance, do you go about tracing the development of "occupance types that exhibit spatial instability" from year to year? How does the operator coordinate operations so as to reduce the time that machines spend in transit and increase the time they spend in the field? Are farm machines really more hazardous on the public road than already sus-

---

[1] Merle Prunty, Jr., "The Census on Multiple-Units and Plantations in the South," The Professional Geographer 8 (September, 1956): 4.

[2] James S. Fisher, "The Modification of Rural Occupance Systems: The Central Georgia Piedmont" (unpublished Ph. D. dissertation, University of Georgia, 1967), p. 153.

[3] James S. Fisher, "Federal Crop Allotment Programs and Responses by Individual Farm Operators," Southeastern Geographer 10 (November, 1970): 47-58.

[4] Charles S. Aiken, "The Fragmented Neoplantation: A New Type of Farm Operation in the Southeast," Southeastern Geographer 11 (April, 1971): 43-51.

pected? We look forward to hearing his answers but will try to suggest some of our own in the meantime.

Whereas these southern geographers were just beginning to think in terms of noncontiguous tracts in the 1950's, the group of midwestern observers was rediscovering what had been known in its area for years. Eisgruber, a Purdue graduate student following a long and wise tradition among agricultural economists at that institution, journeyed to Forest Township, Clinton County, in 1955 to see what had happened to farms and farmers there since the previous Purdue examination ten years earlier.[1] Predictably, he encountered the same sort of tangled pattern of farm operating units that had attracted the attention of Hays and the rest of the 1940 group. Reorganization of holdings in Forest Township by "swapping and consolidation" would have been advantageous, he opined, but the likelihood that it would ever happen seemed slim. Another Purdue student, Kelsey, was also out collecting information from Indiana farmers on distance to outlying tracts; but since this was of marginal relevance to his main research interests, he did practically nothing with the distance data.[2]

Despite their discipline's enduring concern with linkages, geographers expressed little interest in the spatial complexities of midwestern agricultural landholdings prior to the 1960's. That decade, however, brought us the two most ambitious analyses of the noncontiguous tract phenomenon yet to appear. Both studies were set in Minnesota.

Noncontiguous parcels attracted the attention of one of these authors, Smith, because of the traffic they generated on the rural roads he was studying in nine Minnesota townships.[3] Initially, assuming only the need to provide each inhabited rural dwelling with outside access, he decided that perhaps as much as a fourth of all road mileage in these townships could be abandoned thus saving the maintenance costs and returning land at the rate of six acres per mile (of road) to agricultural uses.[4] On the other hand, where would this action

---

[1]Ludwig M. Eisgruber, "Changes in Farm Organization in a Central Indiana Township from 1910 to 1955" (unpublished M.S. thesis, Purdue University, 1957), pp. 14-16.

[2]Myron P. Kelsey, "Economic Effects of Field Renting on Resource Use on Central Indiana Farms" (unpublished Ph.D. dissertation, Purdue University, 1959).

[3]Everett G. Smith, Jr., "Road Functions in a Changing Rural Environment" (unpublished Ph.D. dissertation, University of Minnesota, 1962).

[4]_Ibid._, pp. 50-52.

leave the farmers who had chosen to farm land located away from their farmsteads? Some tracts would still be accessible, but many would no doubt lie on stretches of road suitable for abandonment. Smith went on to explain carefully why there are noncontiguous tracts in the first place, why there will probably be more in the future, and why even though road segments served no farmers in 1960 they might have in earlier years or could in the years to come.[1]

Smith did not attempt to reconstruct the farms that had supposedly made complete road networks necessary in the years preceding 1960. "To determine what constituted their operating units in some previous year," he remarked, "would be a most formidable assignment."[2] Such a task is formidable though hardly impossible. He wisely recognized the disutility of Agricultural Stabilization and Conservation Service records[3] in any sort of serious land tenure search, but he apparently felt an interview campaign such as the type mounted by this author too difficult to undertake. While correctly anticipating the memory bias that one must overcome in this kind of endeavor, Smith incorrectly assumed that persons who have moved away leave unmanageable gaps for the tenure historian.[4]

The Finnish geographer Rikkinen did employ the historical approach in his intensive study of a Finnish-American agricultural community occupying a Minnesota Cutover township in Carleton County, thirty miles west of Duluth.[5] Rikkinen concentrated on changes between 1930 and 1966 in farm size, land ownership, cattle numbers and types, and degree of mechanization. Some of his numerous maps and the insightful comments relating Kalevala Township to outside influences are commendable. There is, however, a serious shortcoming that makes his analysis of farm operating units practically worthless.

Rikkinen chose to take his local data entirely from Carleton County's personal and real property records. Cards had been kept for each forty-acre tract in Carleton showing valuation, taxes assessed, ownership changes, type of land, and types of buildings. Personal property cards told him how many and what kinds of livestock a farmer had; the "true and full value" of his livestock, machinery, and other minor taxables; and sometimes, where the farmer lived. Using only these primary materials and without bothering to check with farmers

---

[1] Ibid.

[2] Ibid., p. 78.

[3] See our discussion of ASCS records in Chapter III below.

[4] Smith, "Road Functions," p. 78.

[5] Kalevi Rikkinen, "Kalevala, Minnesota: Agricultural Geography in Transition," Acta Geographica (Helsinki) 19 (1969): 1-58.

on past or even 1966 rental arrangements, Rikkinen reconstructed what he called "farms." In truth, they are only farm ownership units. We are never sure whether a farmer actually farmed all he owned nor do we know whether he rented any of the 10, 000 or so nonfarmer acres in the township belonging to individuals, the public, and the lumber companies. To make matters worse, Rikkinen totals the farmer-owned acreage, divides by the number of farmers listed in the personal property records, and labels the quotient "average size of farm." In defense of his actions he wrote, "rental practices . . . can only be surmised since the tax books do not reveal such information. . . ."[1]

In summary, noncontiguity research in the United States has been sporadic and rather superficial. With few exceptions, authors have considered the noncontiguous arrangement of farms only at a single instant. No one has really set out to examine thoroughly all facets of the situation as it pertains to the American farmer. In fact, most of the researchers cited above seemingly stumbled onto the farming of dispersed parcels while investigating something else. Smith's Minnesota dissertation is a good case in point.

Evolution and Character of the Present Research Enterprise

As we indicated in the initial pages of this chapter, American farmers have been subjected to a cost-price squeeze for years. To avoid being eliminated from agriculture, the only solutions lying entirely within the agricultural sector available to the individual farmer have been to increase his personal agricultural output (1) by increasing the productivity of the acres already under his control or (2) by increasing the number of acres he farms. Neither solution guarantees success. Increased output requires greater production costs for fertilizers, pesticides, and feed (if the farmer chooses to intensify his livestock efforts). Increasing the number of acres farmed brings the added costs of land ownership or land rent, if--and this is a large if--there is land available to him that he can economically rent or purchase. Farming more acres also increases mechanization investment and, of course, operating charges for this additional machinery.

Farmers' acquisition of more fertilizers, pesticides, farm equipment, feeder livestock, and the like is limited only by their capability to invest. Supply will eventually respond to increases in demand and price. Moreover, all of

---

[1] *Ibid.*, p. 20.

these inputs are mobile. No matter where they are produced, they can be transported to wherever the farmers wish to use them. But farmland is almost completely different. The total supply of land is essentially fixed although the supply of farmable land can be increased by large investments. But land is utterly immobile. It cannot be transported. It can be farmed only where it is. The farmer must go to the land; the process cannot be reversed.

Having an abiding interest in rural America, the author had personally observed the procedures followed by a few families in obtaining more land to farm and knew such efforts must have been ubiquitous and continuous for the past three decades in the United States. Geographical experience suggested, however, that the process of obtaining land through interfarm migration and in-place expansion had probably varied in different locations, circumstances, and times. Seminars and conversations with individuals at The University of Chicago during 1968 led to development of the notion that migration and expansion rates and their attendant problems would be likely to differ according to (1) distance of agricultural areas from major urban centers, (2) value of farmland, (3) quality of farmland, and (4) prevailing type of farming.

Field work commenced in an agricultural township in Will County, Illinois, on the extreme outer fringe of the Chicago metropolitan area, during February, 1969. Planning to use whatever information sources might present themselves, this author set out to examine in minute detail the workings of the rural land tenure system over a lengthy period. In essence, we were concerned with the appearance, expansion, contraction, and disappearance of farm units and with the farm-to-farm migration of farmers seeking different land. As field work proceeded, the research focus shifted somewhat because informants repeatedly brought up additional items of interest. And herein lies the great advantage of personally generating data in the field as opposed to taking it from a prepackaged source. A good indicator of how the research emphasis continued to evolve is the list of questions carried along to help structure the interviews. The first and final lists were substantially different. This author's curiosity about noncontiguity problems and their solutions really crystallized while he was out talking with farmers. Similarly, abandoned farmsteads, another concomitant of farm enlargement, caught our fancy as we spent time among them and their former inhabitants. Farmsteads did not become a major research issue like noncontiguity, but there was throughout the study an underlying awareness of the farmstead (both abandoned and occupied) as an important element in the land turnover process.

Where This Study Fits In

The foregoing has led us to the point where we must establish a niche for this product of our research. What contributions are we hoping to make toward a better understanding of farmer mobility between farms and of farmer movement away from the farmstead to acquire land? How was our approach different? Whereas most authors have been content with only a single year's land tenure protrait, we will be covering a three-decade period. Whereas Malin and others were limited by lack of information on very short interfarm migrations, we will be able to consider all types. Whereas some have depended on one source for operatorship data, we have combined gleanings from courthouse collections, newspaper files, interviews, and other sources to assemble for analysis a relatively complete history of landholding in the three study areas. Whereas some investigators chose to rely on field data that had been gathered by enumerators, we were able to do all the necessary interviewing personally. And finally, whereas many have either overlooked or merely alluded to problems facing the operators of fragmented American farms, we intend to expose these problems and to catalog in detail the efforts being made to solve them.

General Questions

No explicitly formulated hypotheses are being tested here, but questions of a highly general character are being addressed. Some have been on the author's mind from the start of this research. Others came to the forefront during the course of the field work. They are combined in the list below.

1. Why do farmers move?
2. Are midwestern farmers more or less apt to move now than in 1939?
3. How far do families move?
4. Do political, cultural, or topographic boundaries exert an influence on interfarm migrants?
5. Does the age of a farmer reveal anything about the likelihood of his moving?
6. Is the frequent relocation of a farm family detrimental to its community standing?
7. How have farmers learned about farm availability?
8. When during the year do moves usually occur?
9. What proportion of midwestern land is in farmstead tracts? Nonfarmstead tracts? Noncontiguous nonfarmstead tracts?
10. How have farmers acquired these different types of land?
11. Is the distance a farmer is willing to travel to farm an outlier related to its size? Quality? Special circumstances? His age?
12. In terms of extra time, fuel, and wear on equipment, what does it cost to handle noncontiguous land?
13. Besides distance, what other impediments to the movement of machinery and livestock does the dispersed farmer face?

14. How have farmers attempted to reduce the burdens imposed by noncontiguity?
15. What is the current status in the farm equipment industry with respect to implement portability?
16. When did farm tire makers accept the fact that road wear was a serious problem for the farmer?
17. What sorts of assistance has the dispersed farmer been getting from governmental agencies?

As the reader may have observed, the first eight of the above questions pertain to interfarm migration and the balance to farm expansion and the farming of noncontiguous land. In the study as a whole, all of the questions are referred to three selected study areas, each comprising a township--as fully described in the next chapter. Further, these areas are approached throughout the study in two modes: (1) comparison between contemporary and past conditions in a given township and (2) comparison among townships of conditions at a given time.

CHAPTER II

THE STUDY AREAS

A Brief for Using Townships

The midwestern county, although a small and inefficient administrative unit, is much too large for the sort of one-man investigation undertaken here. Fortunately, below the county in the governmental hierarchy there still exists that nineteenth-century anachronism, the civil-political township. Despite the fact that counties and municipalities have usurped its authority, farm people continue to identify with their township and seemed to relish the chance to help immortalize it. Further, the Census Bureau collates data by township, publishing limited results from the decennial population count and offering the balance, plus results from the agricultural census, for a search fee. Outcomes of the annual Assessors' Census are available for townships from the state offices of the Statistical Reporting Service in Illinois, Indiana, and several other states. County tax records frequently appear by township as do those of the federal farm program agency, the Agricultural Stabilization and Conservation Service. County plat atlases commonly display ownership information by township.

If the above justifications for using the civil-political township as a study area have proven insufficient, one can always point to a literature replete with precedents.[1] Perhaps because the marriage is so common in the Midwest, authors have frequently opted to employ civil-political townships that are conterminous with the Congressional-survey townships of the American Rectangu-

---

[1]In the land economics and agricultural economics literature we find the following examples: William ten Haken, "Land Tenure in Walnut Grove Township, Knox County, Illinois," The Journal of Land and Public Utility Economics 4 (February, 1928): 13-24; Leonard A. Salter, Jr., Land Tenure in Process: A Study of Farm Ownership and Tenancy in a Lafayette County Township, Wisconsin Agricultural Experiment Station Research Bulletin 146 (February, 1943); Robert W. Schoeff and Lynn S. Robertson, Agricultural Changes from 1910 to 1945 in a Central Indiana Township, Purdue Agricultural Experiment Station Bulletin 524 (1947); Earl R. Swanson, "Operations Research Techniques," in Methods for Land Economics Research, ed. by W. L. Gibson (Lincoln: University of Nebraska Press, 1966), pp. 191-222. Geographers, too, have used the township as the basic study unit. See, for instance: Smith, "Road Functions"; and Rikkinen, "Agricultural Geography."

lar Land Survey System. On the other hand, their roughly equivalent extent (23, 000 acres or so) and similar shape (square) do simplify comparisons between townships. For this study, three six-mile-square townships were selected. The first in a square civil-political township in Indiana; the second, is a marriage of the civil-political and Congressional-survey in Illinois; and the third, is a Congressional-survey township in Missouri (Figure 1).

Selection Guidelines

Despite the fact that the study areas were deliberately chosen to represent diverse aspects and sections of the agricultural Midwest, all have in common, by design, certain traits other than size and shape. None has experienced an interstate-highway trauma. Virtually all 70, 000 acres belong to private, independent family farmers or their landlords rather than to the public or to large private (agricultural, mining, lumbering) firms. While there is timber standing in all three areas, most is used in some manner by the farmers. Very little land is devoted to uses other than agriculture and transportation. In one township there is no population agglomeration, in another a hamlet, and in the third a hamlet and a village. Large-scale topographic maps, recent plat books,

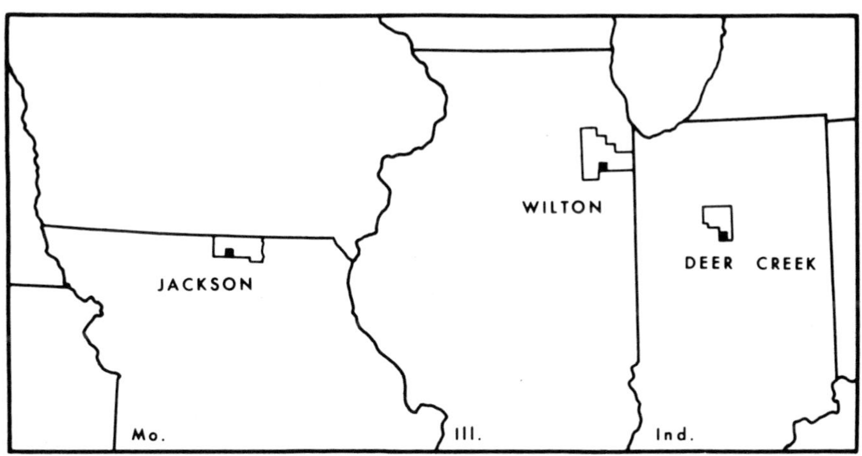

Fig. 1.--The Study Townships

and soil surveys are available for all. In addition, once alerted to each general area the author was able to affirm that:

1. no parochial issues had lately aroused potential informants to affect their overall reception of the project.
2. agricultural advisors were cooperative and willing to sponsor the venture by providing initial introductions to community leaders.
3. local newspaper files and critical courthouse records were intact for the proposed study period and obtainable.

The Choices

Deer Creek Township

In their attempt to unravel and explain land use evolution in a Michigan community, Kaups and Mather note that an historical bench mark is invaluable when seeking to measure change.[1] Unfortunately, it seems that few micro-studies of man's use and organization of space in the past will fit smoothly into a contemporary research plan. Such is the case as one looks through the handful of potentially useful studies for a point of reference from which to begin tracing tenure adjustments and farmer turnover. Some are too recent to provide sufficient historical perspective.[2] Another employs an oddly shaped township.[3] One uses no townships but looks instead at farm layout in three areas consisting of 100 farms each.[4] To different degrees, all these studies ignore parcels lying outside the immediate focus area even though they are just as much a part of the farmers' operations as are those lying inside the area.

Fortunately, however, a Purdue University research team did produce a portrait of land tenure in the late 1930's which is quite suitable to serve as a historical piton for a 1969 restudy. Consisting of agronomists and farm management specialists, the Purdue team assessed the agricultural practices and potential of Deer Creek Township, Cass County, Indiana.[5] John R. Hays, a

---

[1]Matti Kaups and Cotton Mather, "Eben: Thirty Years Later in a Finnish Community in the Upper Peninsula of Michigan," Economic Geography 44 (January, 1968): 57.

[2]Smith, "Road Functions" (1960 data); Eisgruber, "Changes in Farm Organization and Operation" (1955 data).

[3]B. P. Birch, "Farmstead Settlement in the North American Corn Belt," Southampton Research Series in Geography 3 (November, 1966): 30-31.

[4]Headington and Falconer, Size of Farm Units.

[5]Other Indiana townships were also studied, but none fit our needs like Deer Creek.

principal investigator, concentrated on land tenure-land use relationships and reported his findings and recommendations in 1940.[1] Purdue Bulletin 450 is especially useful here because of Hays' concern therein with operator turnover in the period, 1915-39, and with the spatial arrangement of operating units.

If there is a true Corn Belt representative among our three study area choices, it must be Deer Creek (Figure 2). Township farmers in 1969, 1939, and the interim used a substantial portion of the land to produce corn on which to fatten hogs and a sprinkling of beef cattle. Feeder pigs frolicked to maturity in rotation pastures and finished out their short but happy lives loafing in the numerous shady woodlots or well-kept barnlots. According to Hays, Deer Creek was selected by the Purdue experts because it seemed at the time to typify a large section of the Indiana Corn Belt. Musing over the choice later, Hays noted, "it was neither the roughest nor the flattest of areas, neither the best township nor the worst when it came to soil, and it had an intermediate number of livestock." Another township (in southern Indiana) used for comparison with Deer Creek, he felt, probably represented only itself.[2]

Cass County is split by the westward-flowing Wabash River just before it turns southwestward on its way toward Lafayette and eventually the Ohio. Deer Creek Township lies south of the Wabash at the southern edge of the county but at the northern edge of the Wisconsin Tipton Till Plain.[3] The township takes its name from the stream which flows westward across its center into Carroll County where it ultimately empties into the Wabash at Delphi. Deer Creek and other streams have had little success in dissecting the gently undulating ground moraine. Only between the narrow floodplain of the major creek and the nearby edge of the upland does the local relief approach thirty feet. The predominant upland soil association is poorly drained, but few other Cass County soils can

---

[1]John R. Hays, "Land Tenure and Land Use in Selected Areas in Indiana" (unpublished Ph.D. dissertation, Purdue University, 1940); and Relation of Character of Farming Units to Land Management in Two Townships in Indiana, Purdue Agricultural Experiment Station Bulletin 450 (August, 1940). See also: B. R. Hunt, E. C. Young, and Lynn Robertson, Land-Use Adjustments Needed on Farms in Deer Creek Township, Cass County, Indiana, Purdue Agricultural Experiment Station Bulletin 466 (February, 1942). Among the team members conducting interviews with Deer Creek farmers was Earl L. Butz, then a graduate student and later a Secretary of Agriculture under Presidents Nixon and Ford.

[2]Interview with John R. Hays, Oxford, Indiana, March, 1970.

[3]Robert C. Kingsbury, An Atlas of Indiana (Bloomington, Indiana: Department of Geography, Indiana University, 1970), p. 14.

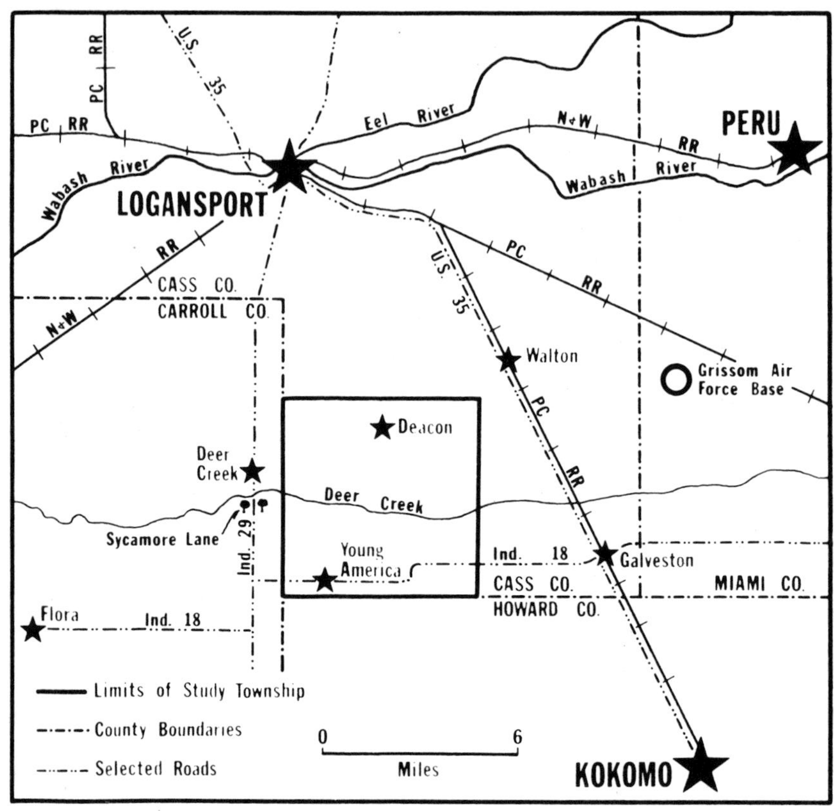

Fig. 2.--Deer Creek and Vicinity

match it in crop yield potential if properly handled.[1] Bottomlands, because they are prone to inundation, generally support only pasture and home gardens if cleared of timber.

Straddling Indiana Highway 18, near the southwest corner of Deer Creek Township, is Young America, the only settlement in the three study areas with occupied, platted streets. Because it is unincorporated and inhabited by fewer than 1,000 persons, the Census Bureau lists its population along with that of the

[1]U.S., Department of Agriculture, Soil Conservation Service, Soil Survey of Cass County, Indiana, by L. R. Smith, W. J. Leighty, D. R. Kunkel, and A. T. Wiancko, Series 1939, No. 24 (Washington: Government Printing Office, 1955), pp. 84-93.

rest of Deer Creek. One unofficial estimate places the 1970 Young America population at 300.[1] From around 1900 the village supported a high school, but low enrollments and projected renovation costs led voters to abandon it in 1963. Students who would have attended the local school now travel by bus to a new rural institution that serves all of southern Cass County from a site near Walton, twelve miles to the northeast. The building in Young America has been demolished, and the land has been converted into a children's equestrian arena.

Handicapped by its unincorporated status, Young America benefits from the activities of a vigorous local men's service organization, the Lions Club. To this group of young and old, farmers and townsmen, the community owes its pleasant five-acre roadside park built in a donated woodlot west of town, streetlights, public restrooms for village loafers (and itinerant researchers), and a spirit of getting things done even though they cannot tax themselves to do it. The Lions finance their benevolence by sponsoring a September corn festival in the streets and a July tractor pulling contest that draws participants from all over the Midwest.

Deer Creek Township lies midway between the cities of Logansport and Kokomo.[2] Logansport, the seat of Cass County, developed at the intersection of the Michigan Road and the Wabash and Erie Canal.[3] It prospered during the late nineteenth and early twentieth centuries as a railroad town astride major routes of the Wabash, St. Louis, and Pacific and the Columbus, Chicago, and Indiana Central.[4] The city has benefited in the twentieth century from its location at the eastern terminus of an important bridge-line railroad around Chicago connecting the Penn Central system to the east and the Santa Fe system to the

---

[1]Richard L. Forstall (ed.), 1971 Commercial Atlas and Marketing Guide (102nd ed.; Chicago: Rand McNally & Co., 1971), p. 209.

[2]It is roughly fifteen miles from Young America to the outskirts of both Logansport and Kokomo.

[3]Kingsbury, Atlas of Indiana, p. 74. The Michigan Road carried north-south traffic from Logansport to and beyond South Bend on the north and Indianapolis on the south. The Wabash and Erie Canal carried traffic over the Maumee-Wabash portage at Fort Wayne east of Logansport, down along the Wabash past Logansport as far as Terre Haute, and then south away from the river to Evansville on the Ohio.

[4]The Wabash, St. Louis, and Pacific Railroad was commonly called the "Wabash" and is now a part of the Norfolk and Western system. The Columbus, Chicago, and Indiana Central became a part of the Pennsylvania Railroad empire and now belongs to the Penn Central.

west.[1] Today, while the Penn Central continues to repair some rolling stock in the old Pennsylvania yards and make crew changes on the Chicago-Cincinnati run here, the city's future depends on the capacity of a miscellany of small manufacturing firms, mainly producing components for other industries, to attract contracts. Logansport's population has stagnated over the past two decades and actually declined 8.9 percent between 1960 and 1970 to 19,255.[2] Perhaps, in view of this, the Chamber of Commerce should stop referring to it as the "City on the Grow."[3]

Kokomo, too, suffered a population loss in the same decade coming down 6.7 percent from over 47,000 to 44,042 in 1970. Unlike Cass County, however, which lost slightly between 1960 and 1970, Howard County gained almost 20 percent.[4] Some of that gain came as Kokomo commuters settled in small towns throughout the county, but much occurred in housing developments just beyond the Kokomo city limits. Twice as large as Logansport, Kokomo is very attractive to the Deer Creek shopper despite his county-seat links with Logansport and probable ambivalence from a distance standpoint. Kokomo also casts an employment shadow across the Cass-Howard boundary into Deer Creek Township. Dominated by several large automobile parts plants, Kokomo's manufacturing employment and value added by manufacturing in 1967 exceeded those of Logansport by about 600 percent.[5] More Deer Creek farm folk now work in Kokomo than work in Logansport; and if Kokomo can maintain its post-World War II economic pace without stumbling, as it has done periodically in the past, that shadow should continue to push northward into Cass County. In spite of Kokomo's obvious strengths, however, Young America High School graduates show no strong affinity for that city over Logansport as a place of residence.[6]

---

[1] Paul Edward Phillips, "The Toledo, Peoria & Western, A Bridge Line Railroad" (unpublished M.S. thesis, Illinois State University, 1967).

[2] U.S., Bureau of the Census, Census of Population: 1970, Vol. I, Characteristics of the Population, Pt. 16, Indiana, p. 56.

[3] Logansport: City on the Grow (Logansport: Logansport Chamber of Commerce, [1969]).

[4] U.S., Census of Population: 1970, Vol. I, Pt. 16, Indiana, p. 57.

[5] U.S., Bureau of the Census, Census of Manufactures: 1967, Vol. III, Area Statistics, Pt. 15, Indiana, pp. 26-27. It is necessary to compare county figures because almost nothing is disclosed for Kokomo in order to protect the large firms.

[6] William G. Wiley, comp., Graduates from Young America High School, 1905-1963 (Flora, Indiana: By the Author, 1969).

Land values in Deer Creek Township continue to reflect only agricultural demand. Real estate developers have been actively competing with farmers for land around Kokomo but mainly to obtain valuable road frontage for house sites. Once the periphery is secure for homes, the parcel interiors are apt to find their way back into the hands of farmers. Young America's areal extent remains stable; but even if it were to expand some, the effects on township land values would be infinitesimal. Subdivisions in Deer Creek Township seem highly unlikely in the twentieth century, so that its residents can expect to enjoy good job opportunities without the bother of urban encroachment.

Wilton Township

Inflated farmland values, abundant and relatively convenient jobs, cross-country petroleum and electricity delivery rights-of-way, vandalism and theft epidemics, and old farmhouses full of hillbillies[1] characterize the second study area and clearly suggest its location on the fringe of a large city. Wilton Township, Will County, Illinois, lies ten miles southeast of Joliet[2] and forty miles from the Chicago Loop at the southern edge of the Chicago Standard Metropolitan Statistical Area in an interstitial zone almost certain to assume progressively a more suburban countenance (Figure 3). An interstate highway, a U.S. highway, an Illinois highway, and the Illinois Central Gulf's north-south trunk run north into Chicago within five miles of Wilton's eastern border. Commuter trains on the Illinois Central Gulf are available fifteen miles from Wilton at the Cook County-Will County line in Richton Park, and growth along this route may encourage extension of electrified service even closer to the focus area. On the west, Wilton is skirted by a four-lane state highway leading north to Joliet, the seat of Will County, and three miles beyond that road by another Chicago arterial interstate. At New Lenox, eleven miles north of Wilton, one can obtain commuter service to the Loop via the Rock Island Railroad.

In spite of its proximity to the aforementioned transportation corridors and a Joliet connection via rough, little-used U.S. Highway 52, Wilton retains its prosperous, rural, agricultural character. In fact, Deer Creek Township

---

[1]Hillbillies, to the permanent residents of Wilton, are low-income in-migrants from the Upper South who often rent large, deteriorating rural houses while they work in area factories.

[2]The population of the Joliet urbanized area in 1970 was 155,500. U.S., Bureau of the Census, Census of Population: 1970, Vol. I, Characteristics of the Population, Pt. 15, Illinois, p. 83.

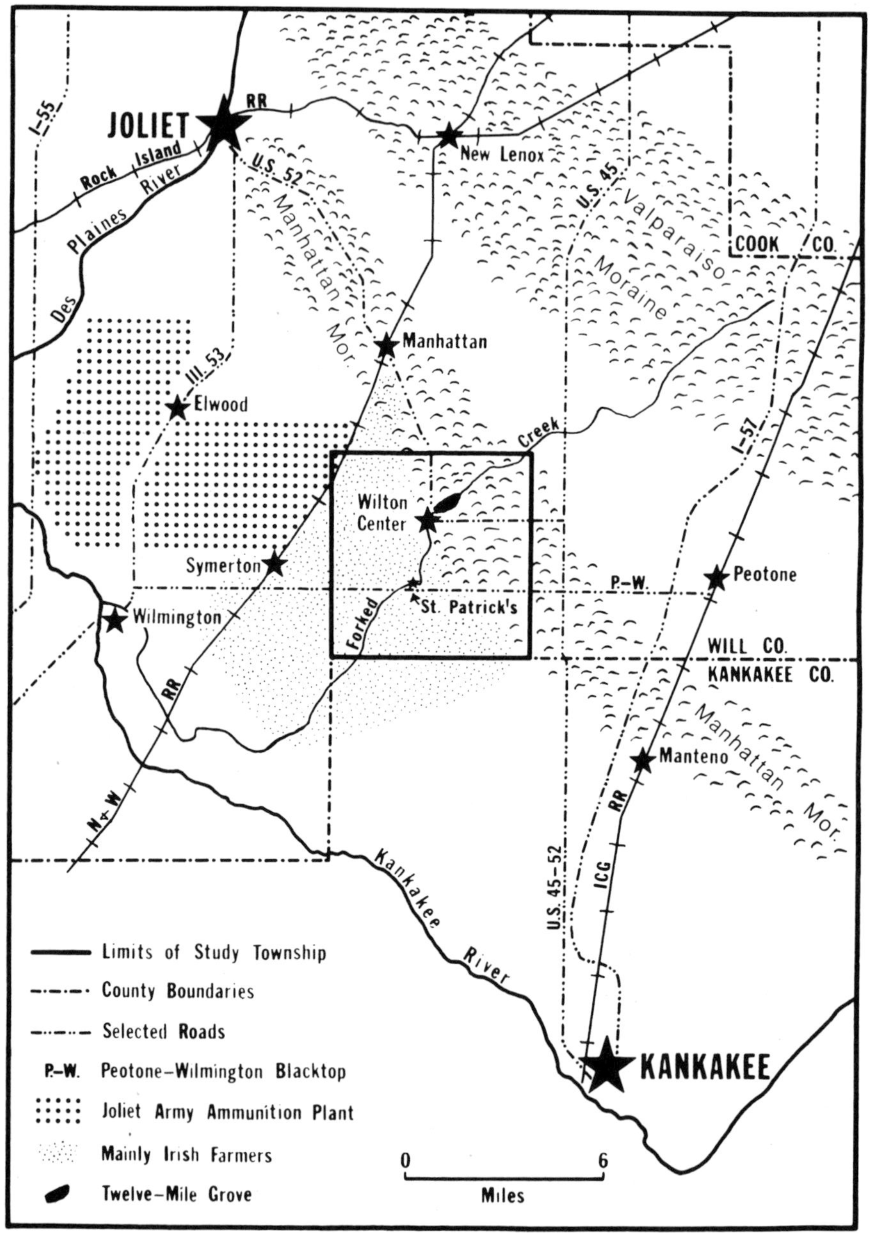

Fig. 3.--Wilton and Vicinity

still counts 352 more inhabitants although Wilton has begun to close the gap (Table 1). Without Young America's 300, Deer Creek's size advantage would have been slight in 1970. Even with Young America, if we accept the Northeast Illinois Planning Commission's forecast of a four-fold increase on the part of Wilton by 1995, Deer Creek will soon lose the population advantage it has held since at least 1940.[1] That same report, however, foresees no manufacturing employment for Wilton during the same period so the township may expect to remain agricultural but to sustain a gradual homeseeker incursion.[2]

TABLE 1

STUDY AREA POPULATIONS: 1940-1970[a]

Township	Census Year			
	1940	1950	1960	1970
Deer Creek	1, 113	996	1, 036	1, 061
Jackson[b]	536	386	280	170
Wilton	609	558	630	709

[a]Sources: U.S., Bureau of the Census, Census of Population: 1960, Vol. I, Characteristics of the Population, Pt. 15, Illinois, p. 27; Pt. 16, Indiana, p. 13; Pt. 27, Missouri, p. 21. Census of Population: 1970, Vol. I, Characteristics of the Population, Pt. 15, Illinois, p. 50; Pt. 16, Indiana, p. 20; Pt. 27, Missouri, p. 33.

[b]Figures for Jackson represent six-sevenths of the number of persons listed as living in the civil-political township proper. For a full explanation see our discussion of the Jackson study area later in this chapter.

Will County is crossed by a series of Wisconsin-age recessional moraines as they loop the southern end of Lake Michigan (Figure 3). Wilton farmers must contend with the rolling Manhattan Moraine throughout the northeastern half of the township and with its extensions northwestward and southeastward

---

[1]Northeastern Illinois Planning Commission, Population, Employment and Land Use Forecasts for Counties and Townships in Northeastern Illinois, NIPC Planning Papers 10 (September, 1968). p. 6.

[2]Ibid., p. 14.

beyond the study area if their operations take them there. Few operators have reached far enough northeast to encounter the much more formidable Valparaiso Moraine. From the outer edge of the Manhattan in Wilton, the land stretches out southwestward across till and outwash to the Kankakee River a few miles above its confluence with the Des Plaines. Drainage paths to the Kankakee are being cut through the Manhattan Moraine by small streams rising on the Valparaiso. Local relief of fifty feet along the largest of these streams, Forked Creek, as it passes through the township's only settlement, Wilton Center,[1] is unequaled elsewhere in the study area. Soils (scarcely acknowledging topographical influences) are just about everywhere high in organic matter, slowly permeable, and moderately to highly productive.[2]

Unlike our once-forested Indiana township, Wilton had a grass cover before settlement and has almost no farm woodlots today. The only appreciable timber stand (perhaps three hundred acres) occurs along Forked Creek north of Wilton Center, especially east of Highway 52. If not for a monumental title tangle, the Twelve-Mile Grove[3] would probably succumb to clearing and improvement, possibly for crops, but more likely to provide homesites for future Wiltonites. Instead, it either lies unfenced and idle or supports browsing livestock.

Wilton lies near, if not upon, the boundary customarily used to separate the American Dairy Belt from the Grand Prairie, a cash-grain section of the Corn Belt.[4] Although near the Chicago market and within reach of milk-shipping points along nearby railroads, farmers south and west of Wilton Center characteristically have preferred either to sell corn directly to their local elevators or to market it through beef cattle. Dairy herds, on the other hand, were once common between Wilton Center and the advancing Chicago fringe.[5] Gradually,

---

[1] Estimated 1970 population was fifty persons. 1971 Commercial Atlas and Marketing Guide, p. 195.

[2] H. L. Wascher, P. T. Veale, and R. T. Odell, Will County Soils, Illinois Agricultural Experiment Station Soil Report 80 (December, 1962).

[3] In largely forested areas like southern Illinois, grasslands were given names; but in prairie areas like central and northern Illinois, groves were singled out by the pioneers. The Twelve-Mile Grove lay roughly that far from Joliet on the route to Kankakee.

[4] For example, see: R. C. Ross and H. C. M. Case, Types of Farming in Illinois: An Analysis of Differences by Areas, University of Illinois Agricultural Experiment Station Bulletin 601 (April, 1956), p. 32.

[5] Loyal Durand, Jr., "Dairy Region of Southeastern Wisconsin and Northeastern Illinois," Economic Geography 16 (October, 1940): 416-28.

the Dairy Belt-Corn Belt distinction is fading in Will County as dairymen defer to age, urbanization, or the inclination to work less on the farm and more elsewhere. Only four Wilton farmers continued to milk seriously in 1969; and of these operations, two are apt to disappear when the present operators retire.

Ethnic clusters persist all over the Chicago area, and Wilton is no exception. Furthermore, it is not improbable that the farm-type boundary crossing southern Will County has reflected more the existence of a cultural seam than it has topographic differences, individual decisionmaking, or manipulations by outside forces.

With the completion of the Illinois and Michigan Canal in 1848, Irish Catholic canal laborers began buying land in southwestern Will County, setting up farms, and urging their countrymen elsewhere to come do the same. By 1860, the Irish had taken much of the land between Wilton Center and Wilmington (on the Kankakee River); and their concentrated numbers had attracted a Catholic mission. Three substantial church buildings were to be constructed (consecutively) by the faithful two miles south and a mile west of the Center on what is now the Peotone-Wilmington Blacktop before St. Patrick's gained parish status in 1905.[1] Since St. Patrick's is under the auspices of the Chicago archdiocese, it has always been relatively easy to find Irish priests to serve there. In fact, according to its first non-Irish priest, assignment to St. Patrick's was regarded not as rustication but as an opportunity to sojourn in the "Garden Spot of the Archdiocese." There a priest enjoyed both an idyllic setting and the goodwill of the Irish parishioners who were anxious and able to provide him with the best.[2] Today, about sixty-five Catholic families, most of them Irish-Americans, continue to maintain the fourth church structure to occupy the spot, a commodious rectory next door, and their own parish priest postponing as long as possible relegation to mission status again. One member feels St. Patrick's would have become a mission back in the late 1920's when the third building burned if it had not been for the rectory. At the moment the priest assigned to St. Patrick's lives in the rectory and serves his parish at night and on weekends but commutes to Joliet daily where he cares for diocesan matters which constitute a part of his responsibility.

---

[1]Information on the history of St. Patrick's gleaned partly from George J. Kuzma, ed., 100th Anniversary of St. Rose of Lima Parish: 1855-1955 (Wilmington, Illinois: St. Rose of Lima Parish, [1955]), pp. 89-93.

[2]Interview with Father George J. Kuzma, Pastor of St. Rose Catholic Church, Wilmington, Illinois, May, 1969.

The Irish have been just as tenacious with their land as with their church. Joined by Irish families worshipping in Manhattan, Wilmington, and Manteno, they own or operate much of the land in the southwestern half of Wilton and adjoining portions of Will County and Kankakee County. Of the few Irish parcels lost to foreclosure during the Great Depression, most found their way back into the hands of other Irishmen in the area. A majority of the Irish farmers produce little except corn and soybeans, but the beef feeders among them appear to be somewhat more prosperous. On the other hand, livestock intrude on the time a man can devote to euchre and gab sessions at the elevator and tavern.

A few families of (pre-Irish) English descent remain on the land near Wilton Center, but most of non-Irish Wilton is farmed by families with a German-Protestant heritage that have sifted in from the old dairying stronghold immediately to the east and northeast of our study area. Dairying, as noted above, is giving way here to less time-consuming agricultural emphases such as grain or livestock fattening. Protestants of all ethnic extractions have successfully supported a federated Baptist-Methodist church at Wilton Center since 1919 when the two local congregations merged their resources. New members affiliate with the denomination of their choice, and ministers are chosen first from one persuasion and then the other. During the late 1950's and early 1960's the congregation built a new edifice at the junction in Wilton Center to replace the Methodist building they had been using. Lack of space for church school and several encounters between the old structure and highway traffic that had failed to negotiate the sharp turn through the hamlet prompted the action.[1] For both the Federated Church and St. Patrick's the future looks promising if the expected influx of new residents can be lured away from lawns, golf courses, and stables on Sundays.

Wilton is only modestly affected by its inhabitants' sharp ethnic, religious, and political differences. Friendships depend more on common age, interests, and locality than on ancestry or beliefs. When the need arises, the two churches cooperate with one another. They maintain adjacent, complementary cemeteries just south of Wilton Center. One respected elderly family even supports both churches--the wife going south to St. Patrick's and the husband east into Wilton Center. On supra-township issues, the Irish traditionally vote Democratic and the Germans and others, Republican; but at the township level, party affiliation means nothing compared to personality and public esteem. For

---

[1] "Wilton Center Church to Celebrate Fifty Years of Federation," _The Peotone Vedette_, April 11, 1969, pp. 1 and 12. Cited hereafter as _Vedette._

many years the township clerk has been an Irishman and the supervisor a German.

One more facet of Wilton's character, the Joliet Army Ammunition Plant, merits mention at this point although more thorough discussion of it will appear later in the study. That the federal government owns a large portion of the western United States and leases it out to nearby ranchers for grazing is a moderately well-known fact. That the same government owns and leases out much more valuable installation land to farmers in various sections of the country for crops and pasture is hardly known at all. Will County's shrinking but still formidable ammunition manufacturing and storage plant occupies sixty-five acres of the extreme northwest corner of Wilton plus twenty thousand odd additional acres between Symerton on the south, Interstate Highway 55 on the west, and Joliet on the north. The "powder plant" has spasmodically offered winter jobs to local farmers and year-round jobs to some whose farms and families permitted or necessitated it. Demand for labor fluctuates violently with the needs of the Department of Defense for bombs, artillery shells, and the like. More consistently since 1944, Wilton and other farmers have been able to bid for and rent portions of the excess plant land as an alternative to the rental or purchase of additional private farmland. A review of Wilton's last three decades is incomplete without an assiduous analysis of the plant and its influence on life around it.

Jackson Township

Traditionally, Iowa and the northern half of Missouri have been considered part of the Corn Belt. All but three Iowa counties and a majority of northern Missouri counties do show corn as their first ranking crop by acreage harvested; and in eight others (one Iowa, seven Missouri) its familiar companion, soybeans, is the leader (Figure 4). There exists, however, a conspicuous knot of counties along the Iowa-Missouri border, away from the Missouri and Mississippi rivers, where hay is the primary crop. Within this rolling island of hay and pasture in Putnam County, Missouri, is Jackson Township, our third study area (Figure 5).

Jackson Township shares with Putnam County and the rest of the interior border country a history of Depression-era farm foreclosures[1] and persistent

---

[1]William G. Murray, Farm Appraisal and Valuation (5th ed.; Ames: Iowa State University Press, 1969), pp. 484-88.

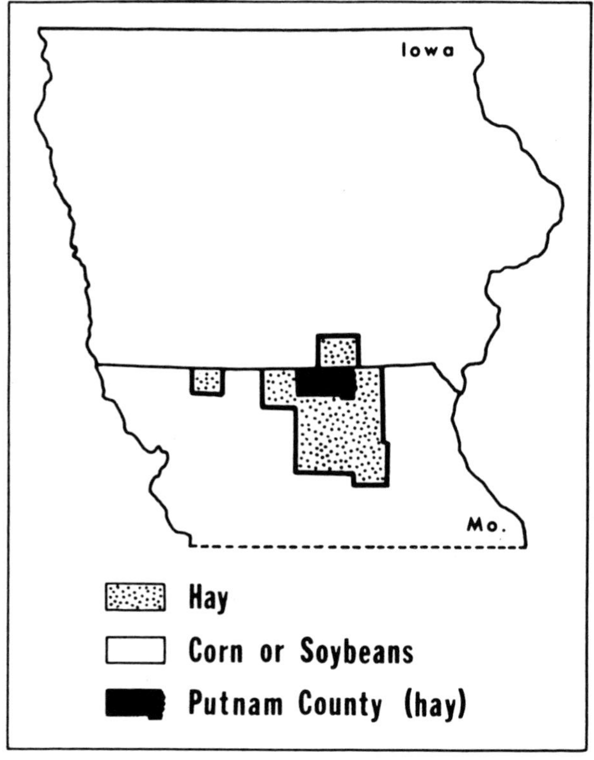

Fig. 4.--First Ranking Crop in Terms of Acres Harvested, by County for Iowa and Northern Missouri: 1969. Source: U.S., Census of Agriculture: 1969.

twentieth-century rural population decline.[1] For the most part, the foreclosures have ceased; but people continue to forsake the area. Total population in a thirty-six-county zone centering on Unionville, the seat of Putnam County, is expected to drop from 442,000 (1967), to 385,000 by 1990.[2] No county, accord-

---

[1]The overall rural population decline in the area is noted in Wilbur Zelinsky, "Changes in the Geographic Pattern of Rural Population in the U.S., 1790-1960," Geographical Review 52 (October, 1962): 511. Village decline is mentioned by John Fraser Hart and Neil E. Salisbury, "Population Change in Middle Western Villages: A Statistical Approach," Annals of the Association of American Geographers 55 (March, 1965): 145.

[2]Peat, Marwick, and Livingston & Co., A Feasibility Study of Recreation

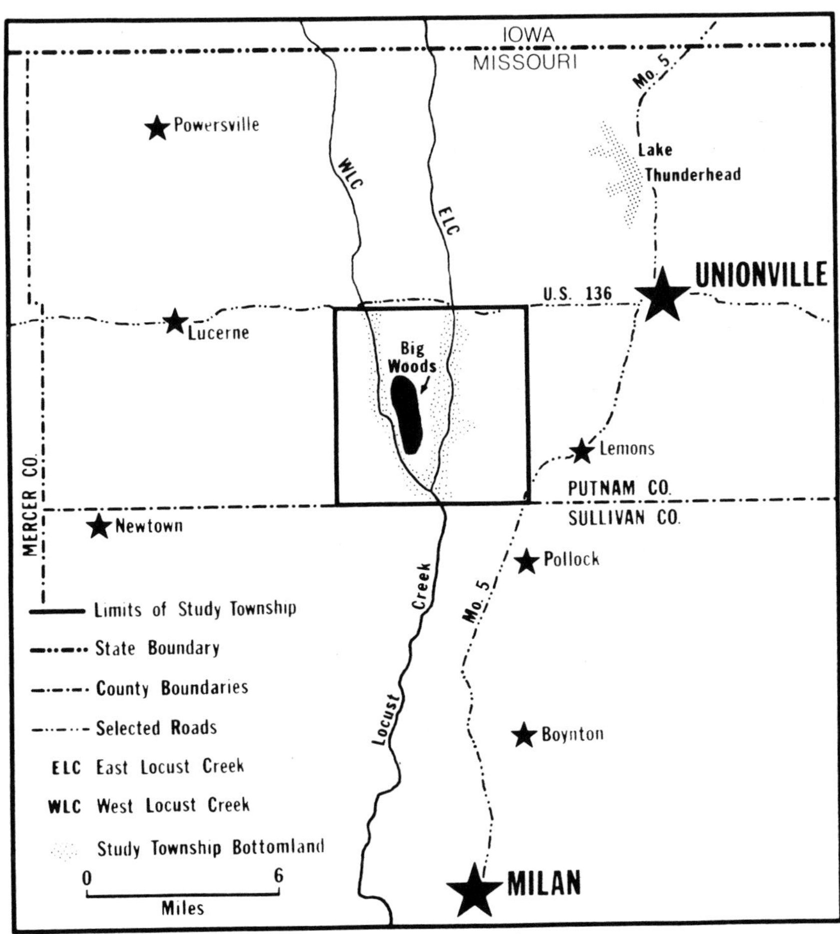

Fig. 5.--Jackson and Vicinity

ing to the forecast, will lose a larger proportion of its inhabitants than Putnam (56 percent). Some of the Putnam decline since its population zenith in 1900 is attributable to the out-migration of families whose men once labored in tiny coal mines punched into bluffs above streams flowing through the eastern part of the county.[1] Jackson's precipitous population loss since 1940, on the other hand, stems directly from the exodus of farm-family offspring or of the families themselves.

As a result of its population problem and low level of living, Putnam appeared on the original list of Economic Development Administration counties issued pursuant to the provisions of the Public Works and Economic Development Act of 1965.[2] Between 1965 and 1969 the county received grants from the EDA totaling more than $600,000. The money was used partly to improve Unionville's water and sewage service and partly to complement a Farmers Home Administration loan of $850,000 from earlier in the 1960's for construction of Lake Thunderhead just north of Unionville.[3] Federal, state, and local funds amounting to more than $100,000 have been combined to build and equip a 2,800 foot airstrip overlooking the lake to serve tourists and local residents.

Encouragement of tourism was a major reason for federal funding of Lake Thunderhead. The lake itself was expected to generate employment and the amenities were to attract new residents and help the county retain those it had. In the late 1960's, when it became obvious that the lake had failed to develop enough even for the Putnam County Lake Association to make payments on the FHA loan, the EDA hired a private consulting firm to prepare a comprehensive plan for the period, 1968-1990. Their report saw an already limited market for day visitors declining as out-migration continued but did envision a good demand for vacation homes along the lakeshore. If properly handled, they felt the complex by 1990 could be yielding a half-million in income each year and

---

and Tourism Development Potential of Lake Thunderhead and Putnam County, Missouri, Prepared for the Economic Development Administration, U.S. Department of Commerce (1969), pp. III.5-III.6.

[1]For a review of coal exploitation in Putnam County see: Richard J. Gentile, Mineral Commodities of Putnam County, Missouri Geological Survey Report of Investigations 29 (July, 1965), pp. 12-27.

[2]"Areas Qualified under Title IV of the Public Works and Economic Development Act of 1965," Economic Development 2 (November-December, 1965): 10.

[3]U.S., Department of Commerce, Economic Development Administration, Annual Report-Fiscal 1970--Jobs for America (Washington: Government Printing Office, 1970), p. 108.

providing jobs for nearly fifty individuals.[1] Early in 1973 the lake (and surrounding acreage) was sold to a Wisconsin man who intends to concentrate on the development of sites for second and retirement homes.[2]

Off-farm jobs are few and pay poorly in this region of surplus labor. A 1968 survey of Putnam County revealed only thirty-one manufacturing employees in the whole county--twenty-one in quarrying and ten in timber processing.[3] Several recent industrial prospects have considered Unionville but decided on other locations.[4] Centerville, Iowa, twenty-two miles north of Unionville and twenty-seven from the nearest corner of Jackson, offers good opportunities if a person is willing daily to drive sixty or more miles, several of which, for the Jacksonian, may be over scantily surfaced township trails. The nearest Missouri cities with employment possibilities similar to Centerville's are both fifty miles from Jackson.

Once the decision to focus attention on Putnam County had been made, selection of a study area from Putnam's eleven civil-political townships proved difficult. Only two of the townships, Richland and Wilson, are coextensive with Congressional-survey townships, and neither seems particularly appropriate for our purposes. Richland is too heavily timbered and in it one begins to encounter eastern Putnam's coal zone where small, part-time farms have severely confused the tenure picture. Wilson would have given a good combination of planosol interfluve and sharply rolling hills, but it includes a portion of Unionville inside its northern margin. For some obscure reason, the next civil-political township to the west, Jackson, embraces an extra tier of six sections to the north. Since Jackson's southern thirty-six sections do correspond with Township 65 North, Range 20 West of the Fifth Principal Meridian and since in all respects Jackson seems representative of the general area, it was decided to examine tenure history in T65N, R20W 5th. P. M. Unless alerted, the reader may assume all Jackson references are to the above-noted Congressional-survey township.

---

[1] Peat, Marwick, Livingston & Co., Lake Thunderhead and Putnam County, Missouri, Chaps. III and VI, passim.

[2] "Sale of Lake Thunderhead Final Today," Unionville Republican, January 24, 1973, p. 1. Cited hereafter as Republican.

[3] Missouri, Division of Commerce and Industrial Development, Unionville, Missouri: Community Profile (1968), p. 3.

[4] "Factory Goes to Brunswick," Republican, January 7, 1970, p. 1; "Factory Did Not Locate in Unionville," Republican, December 20, 1972, p. 1.

Jackson farmers make hay and pasture their beef herds (primarily Angus) on the narrow, contorted remnants of a Kansan till surface;[1] along slopes where local relief approaches 150 feet; and on the flood-prone bottoms of East Locust and West Locust Creeks which slice the township neatly into three north-south sections before linking up at its southern margin. Corn, in Jackson, generally either shares the floodplains with soybeans, pasture, and perhaps some winter wheat and sorghum or serves as the first element in a hillside rotation scheme whose primary goal is meadow rejuvenation. Hogs and dairy cattle are uncommon in southwestern Putnam County but not unknown. Because of distance from processors, milk is sold for manufacturing purposes.

Most of Jackson's woodland acres are fenced and, especially in the winter, utilized by beef cows seeking shelter for themselves and their calves. Woodland, as a proportion of township land, fluctuates as the energetic clear off trees to improve grazing while the uninterested and disabled allow the brush to retake their land. Still extant are a few remnants of the timber that had advanced laterally into the hills from the streambanks prior to settlement. Large tracts which once served as a fuel source for prairie farmers living elsewhere in the county seem to have escaped earnest clearing. Most prominent among those in Jackson is the Big Woods, on the East Locust-West Locust divide, which stretches south two miles from the abandoned Union Grove Church to the edge of the upland overlooking the Locust Creek confluence. Several of the small, idle parcels in the Woods have been incorporated into the domain of the farmer who owns and pastures the bulk of the timber.

Agricultural land use in contemporary Jackson reflects slope constraints more than it did in 1939 and more than it does in the other two study areas where slope is a relatively minor concern. There is, however, room for improved conservation practices since one sometimes finds washy fields harnessed to a corn-soybean sequence and forty-acre segments of floodplain in permanent pasture. To combat upland mismanagement, the local office of the Soil Conservation Service runs full-page admonishments in Unionville's weekly newspaper urging farmers to mend their ways and periodically sponsors field days to accentuate cooperator accomplishments. Recent private drainage projects (with SCS advice) along the Locusts have reduced the flood hazards somewhat and bolstered confidence in row cropping the bottom.[2]

---

[1]One finds little of the broad planosol upland noted in Wilson Township.

[2]To this point most efforts by Jackson landowners have been directed

Farm Size, Land Acquisition, and Land Turnover: An Overview

Size of Farm in Deer Creek, Jackson, and Wilton

The mean, while not necessarily more revealing than the median, is traditionally employed to measure farm acreage. By either standard, farms in the three study townships have roughly doubled in size since 1939 (Table 2). Deer Creek always reported the smallest farms and usually the slowest growth rate but never really lagged far behind the other two.[1] Mean and median disagree as to whether Jackson or Wilton ended up with the larger operations; but in this instance, the median appears to be the better indicator of central tendency. With seven fewer farms in 1969 than Wilton, Jackson had nine farms that exceeded 1, 000 acres compared to only four for Wilton. These nine helped push Jackson's mean above that of Wilton even though the latter had ten more farms totaling between 400 and 999 acres than did Jackson.

Acquisition of Land

Some midwestern farmers own the land they use, some arrange with one or more other owners to rent what they farm, and some do both. Respectively, these farmers are known in official parlance as owner-operators, tenants, and partowners. Owner-operators and partowners have secured title to the acres they own either through purchase or inheritance, so that altogether there are three sources of farmland. The relative importance of these three sources to 1939 and 1969 farm operators is shown in Table 3.

In the Midwest a high incidence of rented land is a good indication of high land costs. Wilton farmland, always comparatively expensive, was selling during the late 1960's for between $600 and $1000 per acre. Even higher prices prevailed just to the north where some suburbanization and much speculation is occurring. Tenancy in Wilton is and has been such a common state of affairs

---

toward straightening out the creeks and clearing of streamside brush to hasten flow at peak periods. In August, 1970, the Missouri Governor's Watershed Advisory Committee recommended priority listing for a comprehensive, head-water-to-mouth watershed improvement plan for Locust Creek. Reconnaissance by the Soil Conservation Service was to begin in 1972. Letter from J. Vernon Martin, Missouri State Conservationist, Columbia, March, 1972. Subsequent dam construction above Jackson will probably serve to attract even more corn and soybeans to the floodplains.

[1] In addition to 1939 and 1969, this author gathered farm size information for five intermediate focus years: 1944, 1949, 1954, 1959, and 1964.

TABLE 2

FARM SIZE BY TOWNSHIP: 1939-1969

Township	Year	Farm Size	
		Median Acreage	Mean Acreage
Deer Creek	1939	126	160
	1944	150	167
	1949	160	171
	1954	160	194
	1959	192	216
	1964	219	247
	1969	252	310
Jackson	1939	162	200
	1944	160	196
	1949	198	224
	1954	200	262
	1959	250	297
	1964	323	371
	1969	353	455
Wilton	1939	178	209
	1944	200	232
	1949	200	252
	1954	239	265
	1959	244	300
	1964	296	342
	1969	380	422

Source: Unless otherwise indicated, all data used to prepare illustrations and tables were collected by this author.

TABLE 3

RENTED LAND, PURCHASED LAND, AND INHERITED LAND AS A PERCENT OF ALL LAND IN OPERATING UNITS, BY TOWNSHIP: 1939 AND 1969

Township	Year	Percent of Township Farmland that Was		
		Rented by Operator	Purchased by Operator	Inherited by Operator
Deer Creek	1939	58	25	17
	1969	69	18	13
Jackson	1939	41	27	32
	1969	33	49	18
Wilton	1939	73	9	18
	1969	77	14	9

that a retiring farmer with no land to bequeathe would consider himself moderately successful if he could just pass on to his son a set of suitable agricultural skills, a set of good machinery, and a set of keys to the farm he was leaving. Jackson farmers, on the other hand, feel an obligation to buy land if at all possible. Fortunately, reasonable prices and the availability of government credit have been making land purchase practical even for the less than prosperous operators. Upland pasture in Jackson during 1969 was bringing no more than $150 per acre and first-class creek bottoms, if for sale, only about $300. Jackson, it should be noted, was just beginning in 1939 to recover from the severe blow dealt it during the early 1930's. A sizable minority of the land still belonged to insurance companies and nonfarm individuals who had obtained it in foreclosure proceedings, so that the proportion of land under lease for 1939 is inflated.

Acres acquired by inheritance are less important now to farmers, relatively speaking, than in 1939. While total acreage in the study areas increased anywhere from 40 to 60 percent during the thirty years,[1] the number of inherited acres increased only slightly in Deer Creek (seventy-seven acres) and actually declined almost a thousand acres in both Wilton and Jackson. For one thing, farmers who were unable in recent years to keep up with the cost-price spiral on farms composed primarily of inherited acres have been forced to accumulate more land through lease and title transactions. Other heirs to farmland have found it easier to make a living outside agriculture and thus their share of the family's land comes up for rent or sale.

Turnover of Land

Farmers and nonfarmers have been receiving and relinquishing rights to midwestern farmland since it was first alienated from the public domain. To help us gauge the amount of land turnover which occurred during our study period, a random sample of fifty-eight quarter-quarter sections was drawn from each of the three townships. Most quarter-quarter sections contain forty acres. We acknowledge but ignore here the fact that, for one reason or another, some have more and some less than that amount. Our sample of fifty-eight repre-

---

[1]Land actually within the thirty-six square miles of the three townships remains at roughly 23, 000 acres each. But since we are also going to consider land lying outside the township limits if linked by operators to land on the inside, there is no ceiling on the total acreage in a given year. Among our three, the maximum recorded was 40, 086 for Wilton in 1969.

sents one-tenth of the total number of quarter-quarter sections in a six-mile-square township. Parcels were rejected if they had been split between different owners or different operators during any of the seven focus years. Rejected parcels were replaced by sufficient acceptable parcels to complete the fifty-eight.

Land moved from owner to owner by means of sale or inheritance with the average sample parcel changing owners slightly more than once during the period, 1939-69 (Table 4). The maximum number of transfers for any one parcel was four and the minimum was zero. In Jackson, where the majority of land was (and is) owned by the operator, ownership changes were more frequent than in our Illinois and Indiana townships where renting is the norm. If a Jackson farmer wants more land, his best bet is to buy it.

TABLE 4

NUMBER OF OWNER AND OPERATOR CHANGES ON SELECTED QUARTER-QUARTER SECTIONS OF FARMLAND, BY TOWNSHIP: 1939-1969[a]

Township	Number of		
	Parcels Sampled	Owner Changes	Operator Changes
Deer Creek	58	72	116
Jackson	58	83	86
Wilton	58	68	120
Totals	174	223	322

[a]We record here only those changes that actually affected a focus year.

Operators turned over more frequently than owners by a substantial margin in Wilton and Deer Creek but just barely did so in Jackson (Table 4). Again, the prevailing methods of land acquisition in the different areas largely accounted for the Wilton-Deer Creek similarity and for the lower number of operator turnovers in Jackson. Farming of a tract to which the farmer holds title is for obvious reasons apt to be looked upon as a more permanent arrangement than one involving merely a lease. A total of twenty-five quarter-quarter sections were farmed throughout the three decades by one (but not the same) person. None of the 174 sample parcels had a different operator every focus year, but four of them were farmed by a different man in six of the seven focus years.

CHAPTER III

GATHERING THE DATA

Had the main objective of this study been only to ascertain the spatial structure of farm units operating in Deer Creek, Jackson, and Wilton during 1969, there would be no reason to devote an entire chapter to field strategy. Data collection under those circumstances would have posed few problems. An analysis of operator mobility and land turnover, however, presupposes access to information about past as well as contemporary tenure patterns. Consequently, a substantial portion of this author's field energy went into the preparation of a detailed series of land tenure situation maps depicting farm ownership and operatorship units in the three study townships for the years 1939, 1944, 1949, 1954, 1959, and 1964. Active participation[1] in local affairs proved impractical, but there was time for casual observation of community life and lengthy conversations with a variety of farm and nonfarm informants.

There were several reasons for selecting 1939 as the initial year. By then the terribly confused tenure picture of the 1930's had begun to clear and the size of farms to creep upward as people left the land and the nation headed toward World War II. The investigation by Hays and his associates in Deer Creek took place at about that time. Construction of the Joliet Arsenal began in the fall of 1940, so that a 1939 starting point affords us a good chance to gauge its effects on Wilton and vicinity. Finally, and obviously, 1939 predates 1969 by exactly three decades.

Land Ownership

An accurate account of changes in land ownership is much easier to obtain than is similar information about land operatorship because people are taxed for owning and not for using a parcel. While counties are anxious to learn of property title transfers,[2] they care not at all about tenant turnover. To con-

---

[1]A sociologist investigating a community will sometimes assume a participatory role in order to understand better what is going on there.

[2]Indiana, for instance, forbids recording a deed of sale before the list of

struct the ownership base maps over which farm operating units were later superimposed, the present author employed commercial plat books, county deed records, and private tract indexes.

Plat Books

A recent plat book exists for each study county and for all but one of the counties immediately adjacent to the townships under scrutiny.[1] Since plat books present a snapshot of landowners (really taxpayers) in a given year and appear only sporadically, it was impossible to employ them as the sole source of ownership data. Despite these drawbacks, plat books served as models and provided clues for construction of the tenure maps. They also saved this author the trouble of having to create cadastral maps from the real estate tax rolls.

Deed Records

Among the American county's most prized possessions are its land transaction records. Using these, the author followed farmland ownership changes within the study townships and in surrounding areas (as necessary) from the 1930's through 1969. From the deed facsimiles that make up the record books one obtains the basic facts about purchase date, location of land, and parties involved. Occasionally, other useful bits of information appear on deeds such as the sale price,[2] addresses of buyer and seller,[3] amendments to reserve usufruct for the seller or his tenant until a specified date,[4] particulars about an

---

real estate taxpayers has been amended to show the new owner's name and address.

[1] Cass-Howard Counties, Indiana: Plat Book and Index of Owners (LaPorte, Indiana: Town and Country Publishing Co., Inc., 1969); Carroll County, Indiana: Triennial Atlas and Plat Book (Rockford, Illinois: Rockford Map Publishers, Inc., 1968); Will County, Illinois: Triennial Atlas and Plat Book (Rockford: Rockford Map Publishers, 1969); Kankakee County, Illinois: Triennial Atlas and Plat Book (Rockford: Rockford Map Publishers, 1967); Putnam County, Missouri: Tri-Annual Plat Book (Marceline, Missouri: Advertising Enterprise, [1966]). The county with no recent plat book is Putnam's southern neighbor, Sullivan.

[2] Actual sale prices seldom appear on deeds, but it is possible between 1932 and 1967 to estimate the price from federal transfer tax stamps. See: Murray, Farm Appraisal and Valuation, pp. 82-87.

[3] Just having the name of the buyer's home county is useful in a mobility study.

[4] For instance, the owner might reserve the right to harvest crops already

earlier transfer of the same parcel, and an indication of what happened to a previous owner whose name fails to appear as a seller.[1] Like plat books, deed records incompletely reflect contract sales and inheritance transfers. Ordinarily, land contracts are not recorded since the deed is kept by the seller until the buyer's obligation is met or enough equity has been built up to justify a loan for the balance due.[2] The number of contracts is fortunately small, and only recently have they become a popular mode of transfer. Devolution of land comes to the attention of the county only if the heir receives and records a deed or if he asks to have the property placed in his name for tax purposes. Many inheritance transfers, however, presented little difficulty because the farm operator typically remained the same through the process.

Tract Indexes

Land transactions are chronologically indexed in the county recorder's office by name of buyer and by name of seller. Occasionally, however, it was desirable to check quickly all transactions that had affected a single tract through the years. An up-to-date abstract of title would, of course, have been the ideal tool in these instances; but abstracts are scattered and closely guarded by their owners or the mortgagee. To remedy this problem, title companies in Putnam and Will counties graciously granted access to the tract indexes which they maintain to aid their work of insuring clear title to land that is changing owners.[3] Regrettably, no tract index exists for Cass County.

Place of Residence

The great majority of midwestern farmers live somewhere on the land they operate. Therefore, any clue to place of residence is also often a clue to what was being farmed. In view of this, every effort was made to locate materi-

---

planted or note the present tenant's name and indicate that he has the right to remain until March 1 of the following year.

[1]When heirs sell land to clear an estate, it is not uncommon to mention in the deed their benefactor's name, the date of his demise, and his relationship to them.

[2]Contract buyers are considered mortgagees in Indiana and entitled to a homestead exemption like others who have received a warranty deed in return for a mortgage. Consequently, land contracts in Indiana are usually recorded.

[3]These were: Ream Abstracts, Real Estate and Insurance in Unionville; and Chicago Title and Trust in Joliet.

als that could be depended upon to confirm that during a focus year a farmer did, indeed, live in the study township, within a certain portion of it, or best of all, on a particular farm. Exposed during the search and profitably employed in the field to determine place of residence were rural directories, local newspapers, and personal property tax assessment lists.

Rural Directories

City directories are routinely exploited by persons studying intraurban migration. For selected midwestern counties, Mio Directory Service Company and Robinson Directories publish somewhat similar rural directories that could have been moderately significant here in the determination of farmers' residences. Unfortunately, only two specimens (both Cass County) were uncovered.[1] Mio's Rural Farm Directory consists of a series of civil-political township maps depicting roads, settlements, and open-country houses. When a house is inhabited, the name of the household head appears in that section and also in the overall county index where his tenure status (owner or renter) is indicated. No hint is ever given as to whether or what the family might have been farming at the time. Robinson uses no maps but alphabetically lists all rural and small-town householders in the county. A Robinson entry typically reveals the names of all residents, ages of children, occupation of the head, precise location of the house,[2] mail route, and telephone exchange.

Local Newspapers

Backfiles of newspapers[3] published near the study areas revealed much about farmers' specific locations and their movements from farm to farm. Two facets of the newspapers proved particularly beneficial: auction notices and personal news columns supplied by country correspondents.

---

[1] Cass County, Indiana: Official Rural Farm Directory, 1969 (Algona, Iowa: Mio Directory Service Company, 1969) and Robinson's 1964 Cass County, Indiana, Rural Directory (n.p.: Robinson Directories, Inc., 1964).

[2] Cass County uses the rural road numbering system explained in A. K. Branham and J. E. Baerwald, "Progress Report on County Road Marking," Proceedings of the 40th Annual Purdue Road School 38 (July, 1954): 117-30.

[3] The Republican, Unionville's weekly, has already been cited. For Wilton we chose to examine another weekly, the Vedette, which is published in the Village of Peotone a few miles to the east. It, too, was cited in Chapter II. A daily published in Logansport, the Pharos-Tribune, provided us with coverage of Deer Creek Township. We cited it fully in the introductory chapter.

Auction Notices

The advertisement placed by the auctioneer in the local paper to alert potential buyers can reveal much to us about the seller. Most farmers whose possessions are offered at auction have decided to close out their operation and quit farming altogether. Not infrequently, since closing out (for one still able to work) is a disgrace of sorts, the advertisement may indicate the circumstances that necessitated this drastic action. Putnam County auctioneers, in particular, went out of their way to assuage the embarrassment of these sellers. Unlike men in the better farming areas of Will and Cass counties, few in Putnam were just quitting. Putnamites instead were:

"taking a job with the Farm Security Administration"
"buying a store"
"buying a motel"
"taking over the Sinclair station"
"going to divinity school"
"drafted"
"volunteering"
"called up"
"going to the army"
"leaving for California"
"going west"
"moving back to Iowa"
"expecting to move soon"
"unable to locate a farm to handle stock."[1]

Sales almost invariably occur at the seller's farmstead because of the difficulty entailed in moving the multitude of items to another farmstead or a central sale barn. Explicit directions are given in the notice to guide to the site those unfamiliar with the area. Thus we learn exactly where the seller last farmed. And since the great majority of closeouts take place between cropping seasons, we also discover when he last farmed there.

Country Correspondent Columns

Republican, Pharos-Tribune, and Vedette editors, knowing that subscribers like to see familiar names in print, have encouraged women throughout their primary circulation regions to write neighborhood gossip columns. Even though the correspondents tend to feature the people best known to them and to under-

---

[1]From auction notices that appeared in the Republican between 1939 and 1969. Country store acquisitions led to several farm sales in Putnam during the three decades. For men anxious to escape farming, a local general store provided an honorable, self-employed alternative that kept them in contact with friends and neighbors. Later, if the store business failed to prosper, they did not feel so badly about accepting other nonfarm work.

report the activities of newcomers, the less neighborly, and those on the neighborhood periphery, they, nevertheless, reveal (through comments on accidents, birthdays, visiting and the like) which families were living in the area at the time and which were moving onto or off of nearby farms. At times a dash of local color even invades the lists of who ate with whom the preceding Sunday.

In 1939, there were thirty-nine country correspondents contributing to the Republican. Of these, seven came from within or along the periphery of Jackson Township. By 1969, the total number of columnists had dropped to fourteen; and only one Jackson correspondent continued to broadcast. Happily for us, coverage of Jackson events was best during the most obscure portion of our three study decades.[1] Wilton and Deer Creek townships never (1939-69) produced more than two columns at the same time, and for 1939 in Wilton there were none. Below are representative items paraphrased from the Republican columns.

December 14, 1938	. . . moved to his mother's farm.
February 22, 1939	. . . and family are moving into the old Greene place.
May 12, 1941	The wolves [coyotes?] are plentiful.
February 23, 1944	. . . is moving machinery into the Hurford farm.
March 8, 1944	. . . and family are leaving soon for Washington state.
March 23, 1949	. . . and . . . were in Lemons Monday morning while the roads were frozen.
November 30, 1949	Folks in the Vernon neighborhood are getting wired for electricity.
May 3, 1950	We were visited with a nice rain last Saturday night.
May 31, 1950	. . . and son, . . . , are farming the Gilworth farm this year.
May 28, 1952	. . . had the mumps and his neighbors planted his corn.
February 18, 1959	. . . has decided to move to town.
August 12, 1964	. . . wrecked an old historic building on his farm recently.

---

[1]In almost every other respect, concrete tenure evidence is poorer for Jackson than for Wilton or Deer Creek.

Personal Property Tax Assessment Lists

Through 1969, personal property was still being taxed in Indiana, Illinois, and Missouri despite vociferous opposition.[1] Persons owning automobiles, trucks, farm machinery, grain, livestock, furniture, and numerous other personal items are assessed by the township assessor at their residence regardless of who holds title to the land.[2] Though possible to escape paying by moving away before the tax is actually collected, everyone living in the township around the first of April is supposedly listed on that year's personal property tax rolls.[3]

Will County's rolls are extant and easily accessible.[4] Each entry gives the name of the person being assessed, his post office, the value of his personal property, and the date the tax was paid. In addition, for 1939, 1944, and 1949, the entry also indicates the number of the small rural school district in which he resided. With the aid of these numbers and the _Atlas of Taxing Units_[5] this author found it possible to narrow the Wilton taxpayer's residential possibilities down from thirty-six square miles to, roughly, four. Then by a process of elimination and a question or two in the community, it was a fairly simple matter to decide which farm he had occupied. If this sort of evidence had existed for all six historical focus years in all three townships, the task of gathering tenure information would have been much easier. Wiltonites in 1954, 1959, and 1964, although taxed uniformly for schools, were still divided into two taxing units because those in the western thirty or so sections belonged to the Manhattan Fire Protection District and the rest did not.

Personal property records for Cass County are incomplete. Hoping to save money, someone in the courthouse decided to reuse the old binders. The 1949 books were selected for experimentation and the pages destroyed at that

---

[1] Illinois has since abolished the personal property tax on individuals. "Legislation Frees Illinois Farmers from Personal Property Tax," _Daily Pantagraph_ (Bloomington, Illinois), July 21, 1972, p. B-12. Cited hereafter as _Pantagraph_.

[2] When a farmer keeps livestock at his residence and at a tract in another township as well, the two assessors may agree to apportion them between the townships for tax purposes.

[3] Official date of assessment varies by state and from one assessor to another.

[4] In the county treasurer's office.

[5] Illinois, Tax Commission, _Atlas of Taxing Units_, Vol. I: _Survey of Local Finance in Illinois_ (Chicago: Illinois Tax Commission, 1939), pp. 83 and 136.

time. Even from the Cass rolls that remain, we learn less about Deer Creek than we did about Wilton because the uniform Deer Creek tax rate relegates all taxpayers to the same alphabetical list. It is possible, however, to eliminate some residential possibilities by noting the mail routes assigned each entry and to distinguish farmers from nonfarmers by depending on the fact that farmers commonly possess much more personal property.

The historian who in 1939 confidently noted "the work of the National Survey of County Archives has served to demonstrate to local custodians the utility of more careful attention to these records"[1] might be somewhat disappointed with the Cass and Putnam custodians. Except for current and very recent years, the books containing the personal and real property lists for Cass County are rotting in a dirt-floored room under a stairway leading to the courthouse basement, apparently due to lack of storage space upstairs. Putnam County, on the other hand, has no storage problem because its tax lists are destroyed after a couple of years so that the binders can be recycled.

Land Operatorship

The present author enjoyed only slight success in his quest for systematically maintained, post-Depression accounts of study-township land operatorship arrangements. County plat books, the Census of Agriculture, and local files of the Agricultural Stabilization and Conservation Service were considered but rejected. Schedules of the Tax Assessors' Annual Farm Census and yearly summaries of land leases in effect at the Joliet Arsenal prepared (and preserved) by the Army Corps of Engineers were valuable during recreation of tenure situations that once existed in Wilton Township. Ultimately, however, even in Wilton, it was necessary to seek operatorship data directly from the local people themselves, either through interviews or mailed questionnaires.

Plat Books

Could there be a more logical place to report current operatorship than in publications which periodically update the ownership situation? Plat book companies may agree, but their problem lies in getting the information since no single depository for operatorship data exists to match the real estate tax records which they use to determine owners. During the late 1960's Rockford Map

---

[1]Everett E. Edwards, "Agricultural Records: Their Nature and Value for Research," Agricultural History 13 (January, 1939): 5.

Publishers, one of the leading plat book producers, offered to include an index of operators if the sponsoring agency in the county would provide the facts. In a few cases, the sponsor assembled volunteers who compiled a list of farmers and their land. A cursory check of Rockford's Illinois plat books for the period, 1965-71, disclosed that of eighty-five counties mapped only eight had such lists.[1] Furthermore, in at least two of the eight cases, a subsequent plat book issued three years later had none.[2] Presumably, the sponsors' efforts had not been matched by the response from users. Will County was not among the eight select Illinois counties, nor were there operator lists in any of the plat books of Putnam and Cass counties.

United States Census of Agriculture

Scholars of nineteenth-century American agricultural matters have access to federal census schedules, but those interested in the present or recent past encounter a firm Census Bureau commitment to maintenance of anonymity for living participants.[3] Even if the schedules for the period, 1939-1969, were available, however, little would be gained by using them as a primary operatorship source. In the first place, all land owned and tended by a farmer is lumped together as if it were in one parcel. An investigator would still need to consult ownership records to learn exactly which parcels belonged to the farmer at the time. Second, a man's rented tracts are separated only by name of landlord. Location of rented tracts is ignored except when they lie in a county other than the one containing the farm headquarters. For focus years prior to 1969, it must be admitted, nevertheless, that the federal schedules would serve nicely as a check on the recollections of informants about their operations in a certain year.

---

[1] Adams, 1967; Boone, 1967; Clay, 1967; Kankakee, 1970; Logan, 1965; LaSalle, 1967; Mason, 1965; Ogle, 1968.

[2] Adams, 1970; Ogle, 1971.

[3] Manuscript census schedules are available for federal censuses taken through 1890. U.S., National Archives, Federal Population Censuses, 1790-1890: A Price List of Microfilm Copies of Schedules (Washington: Government Printing Office, 1969).

Agricultural Stabilization and Conservation Service Records

Although a number of geographers[1] have made extensive use of these readily accessible, locally generated records, for several reasons they proved worthless here as a major source of tenure information. First, the ASCS is concerned above all with the current crop year. After payments are made to farmers for setting aside cropland and officials are satisfied that all requirements have been met, that year's carefully kept records quickly obsolesce. County offices, while obliged to maintain a five-year back file, are also encouraged to conserve space by disposing of the oldest records when a new set goes into storage. Most managers faithfully follow the five-year rule and send anything older to a central collection point for destruction. This author did, however, uncover worksheets describing farming operations in 1959 and 1960 that had been preserved in Putnam County to aid the ASCS County Committee (of elected farmers) in deliberations over contemporary programs still tied to those base years. In addition, all of Putnam's old Conservation Reserve (Soil Bank) contracts dating from the mid-fifties had somehow escaped disposal. Second, corn is the only significantly subsidized midwestern crop; and participation, which is totally voluntary, seldom in a given year exceeds half of a county's farmers.[2] Consequently, about farmers that do not participate in the corn program the ASCS knows nothing unless someone bothers to inform them or they are asked by another county to verify the local compliance of a person who is "going with the government" elsewhere.[3] Third, when a field is share

---

[1] Among them are the following: Walter M. Kollmorgen and George F. Jenks, "Suitcase Farming in Sully County, South Dakota," Annals of the Association of American Geographers 48 (March, 1958): 27-40; and "Sidewalk Farming in Toole County, Montana, and Traill County, North Dakota," Annals of the Association of American Geographers 48 (September, 1958): 209-31; Jack T. Dugan, "A Geographic Analysis of Some Aspects of Land Tenure in Harlan County, Nebraska" (unpublished M.A. thesis, University of Nebraska, 1969); Karl B. Raitz, "The Government Institutionalization of Tobacco Acreage in Wisconsin," Professional Geographer 23 (April, 1971): 123-26; Wayne E. Kiefer, Rush County, Indiana: A Study in Rural Settlement Geography, Geographic Monograph 2 (Bloomington, Indiana: Department of Geography, Indiana University, 1969), p. 90; Merle C. Prunty, "Idle Land Phenomena in Madison County, Georgia," Southeastern Geographer 1 (1961): 39-49; and Fisher, "Federal Crop Allotment Programs and Responses by Individual Farm Operators."

[2] Most of the items cited in the preceding note deal with wheat, tobacco, or cotton areas where farm program participation is more or less compulsory.

[3] When a farmer or landlord agrees to divert land from corn, he must list all other land in which he has an interest including that located in other counties.

rented with the understanding that all crop diversion payments are to go to the owner, the ASCS will deal entirely with the landlord and keep no record of the renter. Even in the current year, therefore, it is absolutely impossible to describe the land tenure situation solely from ASCS files.[1]

Tax Assessors' Annual Farm Census

As they itemize personal property, township assessors in Missouri, Indiana, and Illinois question farmers about the total acreage in their farms and about the crops harvested from the same land the previous year.[2] Replies are sent by each assessor to his county clerk who forwards them for processing to the state office of the Statistical Reporting Service. All three SRS offices then publish an annual state summary by counties,[3] while Indiana and Illinois also issue county reports by townships. Only the Illinois office, unfortunately, has seen fit to preserve the actual schedules. Missouri destroys them almost immediately, and Indiana maintains but a brief back file.[4]

In 1935, the Illinois General Assembly authorized the state's Department of Agriculture in cooperation with the United States Department of Agriculture to "collect, compile, systematize, tabulate, and publish statistical information relating to agriculture."[5] Township assessors were to serve as enumerators

---

His home ASCS office will commonly check with the other offices during the crop year to make sure he is obeying the rules on all his land.

[1]In ways other than those originally anticipated, local ASCS offices and officials in the study counties did, after all, offer much assistance. The Will County manager, William Polley, provided initial introductions and, like William Richardson, the Putnam County manager, spent a good deal of time explaining aspects of the farm programs and discussing local matters. Aerial photographs, an ASCS trademark, were always available for consultation.

[2]Missouri has recently terminated its Farm Census. Officials fear abolition of the personal property tax may force termination of the Census in the states where it persists. These, besides Illinois and Indiana, include: Wyoming, Colorado, Kansas, Nebraska, South Dakota, North Dakota, Minnesota, Iowa, Wisconsin, West Virginia, and North Carolina. Interview with James Kendall, Illinois State Agricultural Statistician, Springfield, June, 1972.

[3]<u>Missouri Farm Facts</u>, <u>Indiana Crops and Livestock</u>, and <u>Illinois Agricultural Statistics</u>.

[4]The Indiana office in West Lafayette refused to permit this author to examine the schedules that it had on hand.

[5]Illinois, <u>Revised Statutes, Annotated</u> (Smith-Hurd), 5-90.

and to be paid fifty cents for each farmer they contacted. The legislators failed to specify what was to be done with the old schedules, but luckily someone deemed them worthy of preservation. Thus it was possible for the present author, with the permission of the Statistical Reporting Service, to view in the Illinois Archives raw Farm Census data collected between 1938 and the present from farmers in Wilton and surrounding townships.

Like the federal agricultural census, the state Farm Census seeks to identify all farm operators living in a township and to learn about their land regardless of its location. Although warned to ignore township land if the operator lived outside their jurisdiction, some Wilton assessors included it anyway.[1] Furthermore, older Wilton assessors, in particular, tended to attribute farmland management to retired fathers rather than to their sons who were really the operators or to give still-active fathers credit for land that sons had gone out and rented on their own. Assessors occasionally even solicited information from the landowner instead of the operator. Because of these and other inconsistencies, the Farm Census cannot stand alone as a source of operatorship data in this study.

It does, nevertheless, provide one of the few <u>written</u> accounts of study-area land operatorship available to us. Circulating between April 1 and July 15, the assessor found most tenure pacts had already been made by the time he arrived. This would not have been the case had he circulated in January.[2] As a township resident, he knew many of the farmers and could remind them of land they might forget to mention. Only the total acreage is given for each man, but odd-sized parcels commonly stand out, thereby practically identifying themselves. The Farm Census is especially helpful when a farmer was sloughing off a little land each year and turning it over to someone else. It also serves nicely to pinpoint the exact years in which men moved between farms of greatly different sizes. Until the late 1960's, names appear in the order of assessor visitation. Consequently, except at the end of the list where the assessor placed the names of people that had eluded him the first time around, we can usually determine a man's farmstead by noting who appears to either side of him on the list and by applying what we already know about the situation from other sources.[3]

---

[1]For Wilton there were six different assessors taking the Farm Census in the seven focus years. Only 1959 and 1964 saw the same man.

[2]Even a July check in northern Missouri would have probably missed a few late haying agreements.

[3]Michael P. Conzen follows the 1860 census taker around a Wisconsin

Joliet Arsenal Lease Summaries

In a research effort such as this there is bound to be some fortune and some misfortune. The annihilation of Putnam's personal property records was unfortunate. On the other hand, finding an almost complete historical account of land leasing by the government at the Joliet Arsenal seems extremely fortunate given the propensity of federal offices to clear out their files regularly.

Agricultural outleasing of unneeded arsenal land began in 1943 or 1944 on a rather haphazard basis.[1] Apparently no list of 1944 lessees exists since responsible officers were hard pressed even then to discover who was farming what. Local memories,[2] therefore, must serve as the primary 1944 source; but by the next focus year, 1949, with the program running smoothly, the overseer (Real Estate Division, Chicago District, U.S. Army Corps of Engineers) had begun to issue its annual Agricultural Lease Program Summary.[3] Using the District's Summary collection, their current-lessee files, and their arsenal land utilization maps[4] it was possible to depict quite accurately the 1949, 1954, 1959, 1964, and 1969 situations involving lessees who at the same time had land in Wilton Township.[5]

---

township using the same clue in his "Spatial Data from Nineteenth Century Manuscript Censuses: A Technique for Rural Settlement and Land Use Analysis," Professional Geographer 21 (September, 1969): 337-43.

[1]Wilton's assessor wrote at the end of his 1944 Farm Census schedules, "The excess acreage reported this year is because some of our farmers have leased some land in the Government Plant which joins the Township of Wilton."

[2]Memories of 1944 are fairly reliable because men often occupied the same arsenal land for some time and only a few Wiltonites were leasing government land then.

[3]The Summary also covers agricultural leasing at other federal installations in Illinois, Minnesota, Wisconsin, and Indiana.

[4]The Corps periodically issues new maps to reflect changes in tract layout and numbers. In addition to maps found in the District Office of the Corps, others were furnished by the JAAP Land Manager, Harold Holz, and by former and current lessees. Both Holz and Corps Real Estate Specialist, Vernon Evans, were extremely kind to this author.

[5]It was always necessary to identify lessees of the JAAP tract which includes sixty-five acres in the extreme northwest corner of Wilton.

Interviews

To the sociologist Junker, field work and interviewing are synonymous.[1] Field data collection by geographers, on the other hand, characteristically involves consultation with a variety of sources, such as those mentioned above, in addition to conversations with informants. Had the present author been able to locate the requisite operatorship data elsewhere, as hoped, the interview in this endeavor would have been relegated to a supporting role because of the difficulties a full-scale interview campaign presents. As it turned out, however, several hundred interviews with farmers, former farmers, and other assorted folk knowledgeable in some facet of local tenure history were eventually necessary before relict tenure situations could be understood.

Since numerous interviewee refusals would have severely crippled the study, measures were taken to encourage cooperation.[2] County extension agents and other local officials were notified and consulted beforehand. At least one official in each county agreed to contact or point out leading farmers in the study townships. An article to describe the project and to emphasize its nongovernmental nature appeared in each area's local newspaper as the field work got under way.[3] In Cass County, field-work commencement happened to coincide with the monthly dinner meeting of the Young America Lions. Invited to attend, this author was introduced and his reason for being in Deer Creek Township briefly explained. Three months later, long after the Pharos-Tribune article had been forgotten by those who had read it, interviewees could favorably recall seeing the author and hearing about the research at the dinner in Young America.

Unlike a survey in which the answers of one person are as acceptable as those of another, the nature of the present project demanded that this author contact specific individuals. Whether functional in 1939, 1954, or 1969, a tenure setup is best remembered and understood by the farm operator who arranged it. With this in mind, every reasonable effort was made to find that farmer, win his confidence, and obtain the pertinent operatorship data. The current (1969)

---

[1] Buford H. Junker, Field Work: An Introduction to the Social Sciences (Chicago: The University of Chicago Press, 1960).

[2] Many informants, particularly in northern Missouri, asked for nothing more than a brief explanation of this author's purpose.

[3] "University Student Conducting Study of Wilton Township," Vedette, March 14, 1969, p. 7; "Writing Thesis on Farm Expansion in Jackson Township," Republican, June 18, 1969, p. 1; "Student to Study Deer Creek Farms," Pharos-Tribune, April 10, 1970, p. 6. A copy of the appropriate clipping was shown to each interviewee.

operator of each township parcel (exceeding five acres) was asked to designate his 1969 holdings, recall when he began farming each parcel, and identify the previous operator of those parcels lying within the thirty-six square miles. The operator was quizzed about land he might have farmed during past years but no longer farmed. In other words, the 1969 operator, together with the present author, tried to recreate his operating unit as of 1939, 1944, and succeeding focus years if any part of it had fallen within the study township. Each farmer was asked how he had originally happened to take up farmland in the township and from where he had come.

Men who no longer farmed study township acres in 1969, but who had at one time during the three decades, fall into several groups. Some are still farming elsewhere, many are working at a different trade, others are retired, and quite a number are dead. Common to all four groups, including the last, is the tendency to reside (or repose) within a few miles of the three townships. Most of those living within twenty-five miles were eventually contacted and convinced to recall the extent of their holdings in the appropriate focus years, details of their movements into and out of the township's agricultural picture, and names of others who farmed the various parcels before and after them. To previous operators now living beyond a reasonable driving distance letters of inquiry were mailed. Each letter explained the study to the addressee, asked specific questions about his tenure experiences, and supplied dates (if available) for his reference.[1] A copy of the newspaper article written by this author to explain the study accompanied the letter; and in a few cases, a note from a close friend or relative went along as further reassurance of the request's legitimacy.

When, because of death, senility, insanity, refusal, or whatever, a former farmer was unavailable to answer questions, the author attempted to interview those persons, especially relatives, with whom the man had been closely associated. Children, wives, brothers, and parents that had once labored on the land beside the farmer stood a much better chance of straightening out his tenure history than would an elder statesman or township assessor. Exceptions to the relative rule were allowed when an unrelated respondent and the farmer under consideration had regularly traded help or farmed across the property line from one another. Only as a last resort were facts concerning an individual randomly sought.

---

[1]Dates of familiar personal or community events drawn from newspapers, deed records, vital records, tombstone inscriptions, and other sources were used extensively for prompting all informants on tenure matters.

Interviewees were seldom able to rely for support on documents kept in the home. Most farm tenure arrangements are oral, and the few written leases never seemed available.[1] One farmer had purposely saved a lease from 1954 because of the amazing number of stipulations to which the landlord had been able to secure his acquiesence, but this was an exception. Farmers were sometimes willing to consult tax and other records, especially if inability to answer from memory bothered them enough. Diaries, an important source for agricultural historians, were practically non-existent among the people contacted for this study. Repeatedly, the best supplementary source around the home for the farmer turned out to be the homemaker. Whether actively participating in the interview or going about their daily activities, farm wives were often called upon by husbands to resolve tenure dilemmas. "I know I farmed that wet eighty for a couple of years during the war," he might say, "but I'm not right sure when. Let me ask my wife. She's better at dates than I am."

Once the land tenure questions were behind them, men could relax, talk about the community, and offer their views on farm expansion and farmer turnover. Rather than attempt to obtain concrete answers to a set of prearranged queries, this author found it advisable to encourage discussion of topics having special relevance for the interviewee. As expected, the interviews contributed much in the way of fresh ideas.

---

[1] Written leases are scarce, but recorded written leases are even more scarce since recording costs a few dollars and is unnecessary from a legal standpoint.

CHAPTER IV

INTERFARM MIGRATION

The United States has one of the world's most residentially restless populations. Every year a fifth of all Americans move to another home, and by the end of five years roughly one-half will have relocated.[1] Today, among the least restless members of the population are the farmers. Of the slightly more than forty-six million American males employed in 1970 at an occupation other than farming or managing a farm, 19.7 million or 43 percent had been living in a different house five years earlier.[2] For the farmer-farm manager group, the corresponding figures for the same period were 1,345,000; 235,000; and 17 percent.[3] We begin our discussion of the midwestern farmer's spatial behavior by considering his movement from farm to farm. In this chapter we will (1) examine portions of the data collected in our three townships on farmstead use and farmer mobility, (2) elaborate on several of the reasons for farmer movement in these areas during the last three decades, and finally (3) show that moving was an accepted way of life for countless hardworking farm families.

Data Gathered in Deer Creek, Jackson, and Wilton

On Farmstead Usage

Farmsteads[4] are far more durable than the farmers who use them. The proportion of 1969 farmers who were also farming in the same township thirty years before is less than half as great as the proportion of 1969 farmsteads that were being used by farmers in 1939 (Table 5). Even in Jackson, where the pas-

---

[1] James W. Simmons, "Changing Residence in the City: A Review of Intra-urban Mobility," Geographical Review 58 (October, 1968): 622.

[2] U.S., Bureau of the Census, United States Census of Population: 1970, PC(2)-2B, Mobility for States and the Nation, p. 26.

[3] Ibid.

[4] By "farmstead" we mean the site where the house and the outbuildings are located. Thus the farmstead could outlast not only several generations of farmers but also more than one farmhouse.

TABLE 5

FARMER AND FARMSTEAD DURABILITY, BY TOWNSHIP: 1939-1969[a]

Township	Proportion of 1969	
	Farmers Who Were Farming in the Township in 1939	Farmsteads Also Being Used As Such in 1939
Deer Creek	23	76
Jackson	35	83
Wilton	28	83

[a]Here and throughout Chapter IV the tabular data pertain only to farmers, farmsteads, and land actually in the study townships proper (unless otherwise noted).

ture and hay economy enables men[1] to stay on the land longer than they do in Deer Creek and Wilton, only 35 percent of the 1969 group were in the township three decades earlier.[2] As farm operators decline in number, the farmsteads with better buildings or better locations are singled out and reused while those found lacking are abandoned or converted to nonfarm use.[3]

Most farmsteads were used in several of the seven focus years (Table 6). In fact, continuous use was more common than any of the other six possibilities.[4] None of the 448 farmsteads, however, had as many as six different users and only one (in Deer Creek) recorded five. There was a strong tendency for farmsteads to serve but a single user, and of these one-family farmsteads, most numerous were those used in just one focus year or for the entire three decades.

---

[1]There are usually a couple of middle-age or elderly female farm operators in Jackson each year. Mainly widows of farmers, they care for the stock and hire their hay baled. When a farmer leaves behind a widow who continues to manage the place, we assume there has been no change in operators.

[2]A fifth of the 1969 Jackson farmers were on the same farmstead in 1939.

[3]For one geographer's view on the subject see: Harry Franklin Lane, "Abandoned Houses in Oconee County, Georgia: Indices of Changing Land Use" (unpublished M.A. thesis, University of Georgia, 1963).

[4]Since we are not concerned here with combinations of years, the six possibilities (besides continuous use) are simply 1, 2, 3, 4, 5, or 6 foci.

TABLE 6

FARMSTEAD USE AND LENGTH OF TENURE FOR THE LONGEST USER, BY TOWNSHIP: 1939-1969[a]

Number of Focus Years Used as a Base

Deer Creek		1	2	3	4	5	6	7	
Number of Focus Years For the Longest User	1	17	5	3	3	1	0	0	
	2		12	12	9	1	4	0	
	3			13	7	2	9	3	
	4				15	9	6	11	
	5					12	2	8	
	6						11	5	
	7							16	
Totals		17	17	28	34	25	32	43	= 196 Farmsteads

Jackson		1	2	3	4	5	6	7	
Number of Focus Years For the Longest User	1	10	7	1	4	0	0	0	
	2		10	9	3	5	2	0	
	3			6	1	4	4	3	
	4				7	6	4	9	
	5					6	2	6	
	6						5	6	
	7							10	
Totals		10	17	16	15	21	17	34	= 130 Farmsteads

Wilton		1	2	3	4	5	6	7	
Number of Focus Years For the Longest User	1	8	1	1	1	0	0	0	
	2		6	9	3	3	2	2	
	3			6	4	3	2	5	
	4				6	3	7	13	
	5					4	4	6	
	6						6	5	
	7							12	
Totals		8	7	16	14	13	21	43	= 122 Farmsteads

[a]Table 6 is not as difficult to comprehend as it may seem at first glance. Across the top of each township's matrix are the number of focus years in which the farmsteads could have served as a base of farm operations. In Deer Creek, for instance, 17 farmsteads were used only 1 year, 17 were used for 2 years, 28 for 3, 34 for 4, 25 for 5, 32 for 6, and 43 out of 196 for the entire 7. Along the left of each matrix are the numbers 1 through 7 which correspond to the years a farmstead was used by its most persistent user during the study period. Again with Deer Creek as an example, let us consider the 25 farmsteads used for 5 focus years. Of these 25, 1 saw no operator stay for more than 1 focus year, 1 had a maximum operator tenure of 2 focus years, 2 had a 3-year maximum, 9 had a 4-year maximum, and 12 had the same operator throughout the period.

Also apparent, but difficult to explain, was the popularity of tenure periods for the longest user lasting six to fourteen years (covering two foci) or sixteen to twenty-four years (four foci).

On Farmer Turnover

In the foregoing discussion of farmsteads and their use we have intentionally avoided the term "occupant" since it is possible to use a place as one's farmstead without actually living on it. Among the 1969 users of Wilton's sixty active farmsteads, for instance, were three sidewalk farmers, i.e., farmers residing in nearby towns. These men and others like them in Deer Creek and Jackson will not be considered in the following paragraphs. On the other hand, farmsteads that housed two different farmers, such as a father and son with clearly distinct operations, will contribute twice to our turnover statistics.

We will be using the verbs "migrate" and "move" along with their derivatives more or less interchangeably here and placing no restrictions regarding the distance to be traveled or the political boundaries to be breached *en route*. The Census Bureau, on the other hand, differentiates between "migrants," who cross a county boundary during the course of their journey, and "movers," who merely change habitations within the same county.[1] In census usage, therefore, less than one-quarter of the farm-to-farm relocations in our three townships constituted true migrations (Table 7). Of the eighteen situations depicted in Table 7 only one (Deer Creek in 1944) shows more migrants than movers. Jackson Township's new arrivals on farms, in particular, were apt to have begun their treks within Putnam County. A bit later in this chapter we will discuss the possibility of the county boundary as a barrier to interfarm movement.

Less Movement Today than in the Past

Farmers are considerably less prone to move today than they were in the early 1940's (Table 8). Since then there has been an appreciable rise in the proportion of those who were on the same farm five years earlier, at the expense of the other categories--those who moved within the township, those who moved into the township, and those new to farming.[2] For Wilton the increase from

---

[1] *United States Census of Population: 1970*, *Mobility for States and the Nation*, p. 1.

[2] Most members of this last group had not farmed on their own before, but a few had and were returning to agriculture.

TABLE 7

MIGRANTS[a] AND MOVERS[b] TO FARMS IN DEER CREEK, JACKSON, AND WILTON: 1944-1969

Township	Focus Year	Number of Migrants	Number of Movers	
Deer Creek	1944	15	14	
	1949	7	11	
	1954	3	14	
	1959	3	8	
	1964	3	5	
	1969	2	4	
		33	56	= 89 Newcomers
Jackson	1944	3	26	
	1949	2	17	
	1954	0	18	
	1959	0	5	
	1964	1	8	
	1969	0	1	
		6	75	= 81 Newcomers
Wilton	1944	3	22	
	1949	4	16	
	1954	2	8	
	1959	2	10	
	1964	2	4	
	1969	2	4	
		15	64	= 79 Newcomers

[a]A migrant began his journey in a different county.

[b]A mover began his journey in the same county as the township. See the text for a full explanation.

64 percent stability in 1944 to 88 percent in 1969 was without an interruption. Jackson, too, saw a steady climb, with the single exception of a minor reversal between 1959 and 1964. Deer Creek figures drop slightly from the initial 69 percent stability recorded for 1944 to the 65 percent for 1949. In large measure this was attributable to the postwar rush by veterans, and by young men who had stayed with their parents during the war, to the small corn-hog operations that dotted the north-central Indiana landscape. A notable 23 percent of the 1949 Deer Creek farmers had not been farming immediately before occupying that place.

TABLE 8

FARMER MOBILITY AND STABILITY, BY TOWNSHIP: 1939-1969

Township	Focus Year	Percent of Farmers in Specified Focus Years That			
		Were at Same Farmstead Five Years Earlier	Had Moved There Since Last Focus From Within Township	Had Moved There Since Last Focus From Outside Township	Were Not Farming Right Before Arriving
Deer Creek	1944	69[a]	5	15	12
	1949	65	3	9	23
	1954	74	8	5	13
	1959	76	6	3	15
	1964	83	3	5	9
	1969	82	3	3	11
Jackson	1944	66	18	8	8
	1949	66	8	12	14
	1954	69	16	6	10
	1959	78	6	-	16
	1964	77	7	8	8
	1969	88	2	-	10
Wilton	1944	64	7	17	12
	1949	65	10	10	15
	1954	73	5	7	16
	1959	77	8	8	8
	1964	86	4	4	5
	1969	88	4	7	2

[a]The sum of this and other lines may not equal exactly 100 percent because of rounding.

Direction of Movement

Intraurban migration patterns have received a good deal of attention from American geographers lately. Of common concern to several has been the matter of where moves begin and end with respect to the central city and other significant components of the metropolitan area.[1] A similar body of literature

[1]John S. Adams, "Directional Bias in Intra-Urban Migration," Economic Geography 45 (October, 1969): 302-23; Ronald R. Boyce, "Residential Mobility and its Implications for Urban Spatial Change," Proceedings of the Association of American Geographers 1 (1969): 23; Stephen M. Golant, The Residential Location and Spatial Behavior of the Elderly: A Canadian Example (Chicago: Department of Geography, The University of Chicago, 1972), pp. 103-23; Eric G. Moore, "The Nature of Intra-Urban Migration and Some Relevant Research Strategies," Proceedings of the Association of American Geographers 1 (1969): 113-16; Simmons, "Changing Residence in the City," pp. 622-51.

does not exist, as far as the present author knows, to enlighten us about the directional characteristics of interfarm migrations in the United States.

Within the Townships

For the most part, internal movers had in common only the fact that they originated and ended their treks within the confines of their respective townships (Figure 6).[1] In Deer Creek Township the presence of Young America had, as expected, little effect on farm-to-farm shifts. Deer Creek (the stream) may have served as a minor barrier to internal relocation, but the evidence is inconclusive. Of the thirty-four internal Deer Creek movers, eight crossed the creek. For many years it served as the boundary separating the Deacon School District from that of Young America. Until the schools were closed in the 1960's, a rivalry of sorts existed between the two districts so it is reasonable to assume these feelings may have colored a tenure decision now and then.

Internal movement peaked earliest in Jackson (1944), compared to 1949 for Wilton and 1954 for Deer Creek, when a total of twenty families who had been in the township before found themselves on new farmstead tracts. To some extent this wartime mobility in Jackson reflected adjustments by draft-age farmers hoping to diminish their changes of being conscribed. Also implicated, however, must be the insurance companies and other unwilling landowners whose policies toward land rental and disposal were still influencing tenure decisions a decade after the worst days of the Depression.[2] School district boundaries and waterways were checked in Jackson for evidence of influence on internal flows of farm families. Although the two Locust creeks had little barrier effect, a few sizable but flood-prone farms along them did have relatively brisk turnover rates until owner-operators took charge and finally began instituting drainage control measures. Through 1949 Jackson had four elementary school districts. Of the twenty-nine internal shifts for 1944 and 1949, just six movers penetrated far enough into another Jackson district to have forced a change of school for children in the family. It is difficult to say whether or not this represented restraint on the part of farmers to avoid school disruptions.

During the forties, Wilton embraced all or part of ten different elemen-

---

[1]Occasionally, routes are obliterated because farmers retraced their path, or that of someone else.

[2]See our discussion below of the role played by creditor landlords in the interfarm migration picture.

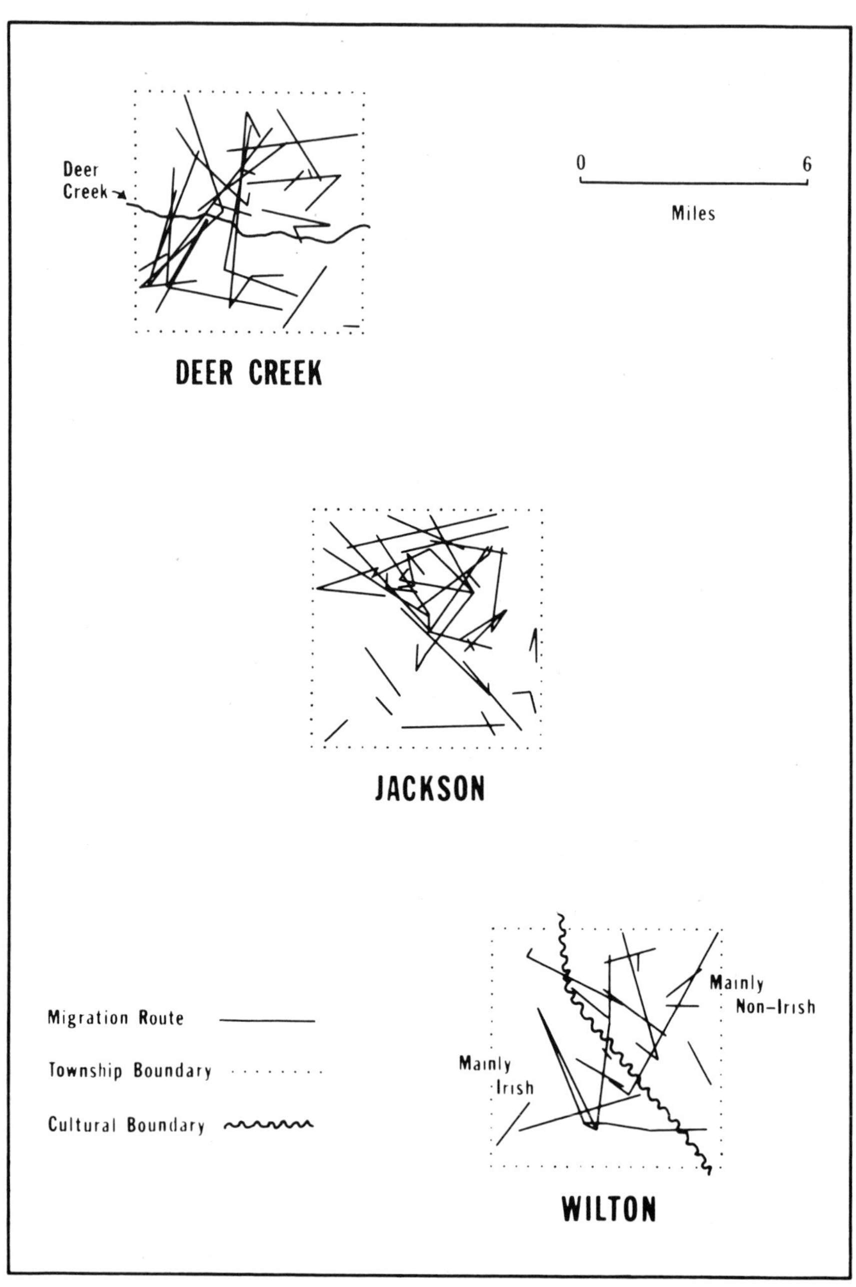

Fig. 6.--Internal Migration of Farmers, by Township: 1939-1969. Origins and destinations are not differentiated. Routes are cross-country.

tary school districts. Since these were necessarily smaller than those in Deer Creek and Jackson, few internal movers failed to cross from one to another.[1] Running through Wilton, however, is a distinctive cultural boundary. It separates, the reader will recall, a predominantly Irish south and west from the German (and other non-Irish) section to the northeast (see also Figure 3). Only 14 percent of internal moves made to Wilton farms took the farm family across this line. While topography and type of farming probably contributed to the preferences expressed by these local movers, kinship and friendship ties seem to have been dominant forces.

Into the Townships

Each township under consideration here has a common border with at least one neighboring county; Deer Creek borders two. Could the position of a township on the county line affect the sources of in-migrants? We noted in Chapter II how rural people generally have a strong attachment to their minor civil division. Similarly, they are often bound because of family, school, shopping,[2] courthouse, and other ties to their county of residence and to its seat of government. In Figure 7 we have depicted the migration routes of farmers who came into our three townships for the focus years beginning with 1944. If the farm of origin was more than a mile from the township line, the initial portion of the crow-flight journey has been omitted.

For Wilton and Jackson there should be no question whether the county lines served as barriers to incoming farmers. In-migrants to Wilton farms were twice as likely to cross the western, northern, or eastern borders as they were to come north from Kankakee County. Ratios in Jackson favoring western, northern, and eastern origins were on the order of 3:1. We might hypothesize that, in the case of Wilton, something other than a political boundary could have influenced in-migration routes; but for Jackson such arguments would vaporize. There is no evidence whatsoever of sharp religious, political-affiliation, or type-of-farming differences in that area which could prejudice farmers' decisions about where to move. Furthermore, the grain of the land in that portion of northern Missouri is clearly north-south. Topographically, therefore, Jackson

---

[1] For over two decades, Wilton's public school children have attended the consolidated school at Wilton Center through grade eight. Some Roman Catholic families though send their offspring to parochial schools in Manhattan or Wilmington.

[2] See: Joel R. Malnick, "Rural County Boundaries as Possible Barriers to Retail Interaction" (unpublished M.A. thesis, Indiana University, 1968).

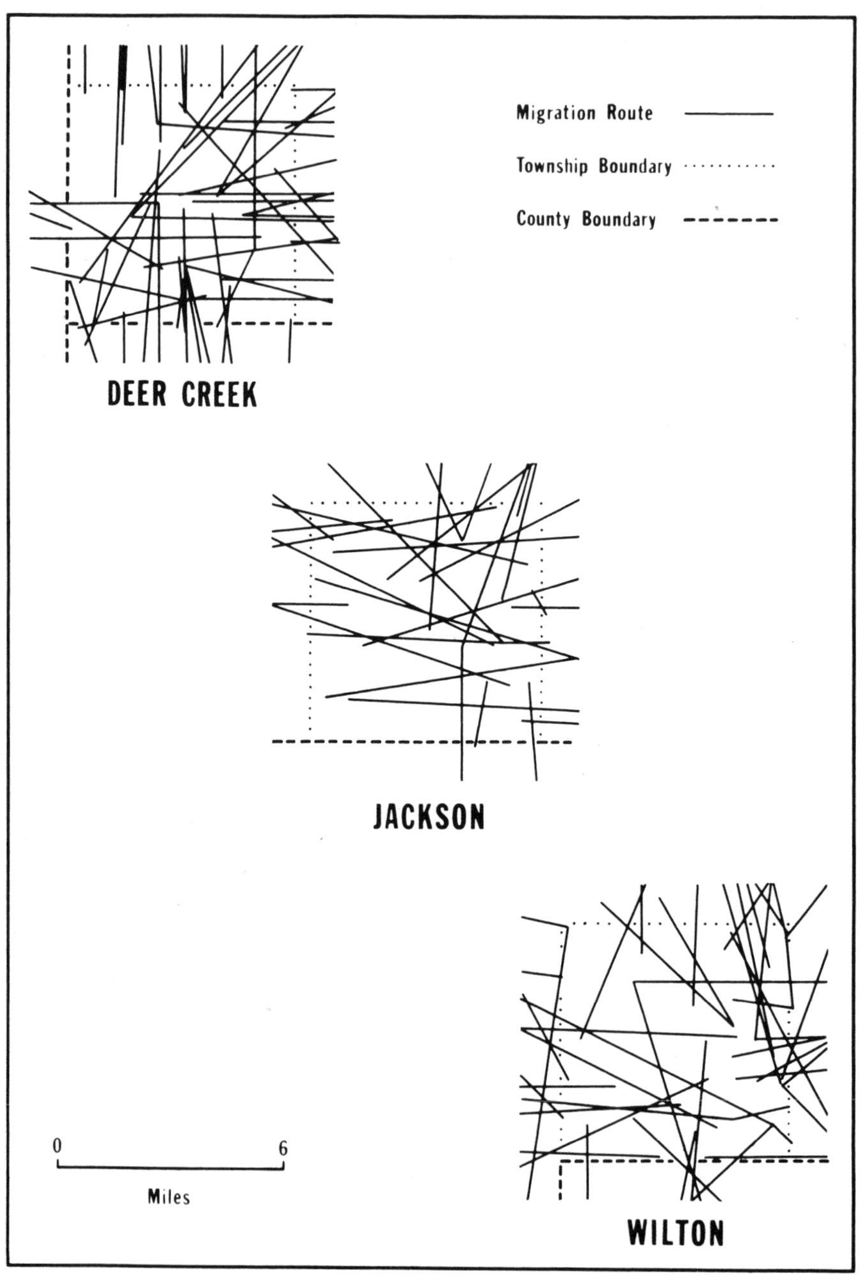

Fig. 7.--In-migration of Farmers, by Township: 1939-1969. Routes are cross-country. Origins are shown only if located within one mile of the township.

Township in Putnam County has more in common with Jackson Township of Sullivan County just to the south than it does with other Putnam County townships to the east and west.

There is less indication of a migration barrier between Deer Creek and its two adjacent counties. In fact, Deer Creek makes a poor subject for this type of analysis because of its peninsular location in southern Cass County. True, the leading source of in-migrants was the area to the east; but many of the eighteen who entered from that direction came from beyond the five-mile wide strip which separates Deer Creek from Miami County. Why, we might ask, did our Indiana township receive twice as many new farmers from Howard County and points south as from Carroll County and points west? Kokomo, the seat of Howard County, could be the key. Working part time in a Kokomo factory or having access to the counsel of someone who did, a Howard County farmer probably stood a better chance of hearing about available land in Deer Creek or of meeting landowners from there than a man would who was farming full time out in Carroll County. On several occasions, while visiting with Deer Creek farm operators, this author detected old links such as these with Kokomo's industrial sector.

Distance Moved

Most urban migrants, according to Simmons, make moves that are "short, within familiar territory, and reflect both satisfaction with the neighborhood and location with respect to the urban structure."[1] Wakeley found the same to be true of movers within his rural population sample: "the distances traveled are usually small even though the number of movers is relatively large."[2] Table 9 summarizes information on the journeys of those who were on new Deer Creek, Jackson, and Wilton farms for 1944 and 1969. Short distance moves were obviously the rule in these areas, too. Because of the large contingent of internal movers, Jackson's 1944 median distance was particularly small. On the other hand, Jackson did have the single most mobile mover for 1944, a man who left his Idaho farm to make an eleven-hundred mile trek back to a place near the southeastern corner of the township.[3] We separated the 1944 newcom-

---

[1] Simmons, "Changing Residence in the City," p. 640.

[2] Wakeley, Differential Mobility, p. 298.

[3] This was easily the longest distance traversed by a farmer headed for one of our three townships throughout the study period.

TABLE 9

MEDIAN DISTANCE TRAVELED TO FARMS BY NEWCOMERS, BY TOWNSHIP: 1944 AND 1969

Township	Year	Median Distance in Miles[a] Traveled by		
		All Newcomers	Renting Newcomers	Buying Newcomers
Deer Creek	1944	7	8.5	5.5
	1969	4	b	b
Jackson	1944	2	2	3
	1969	b	b	b
Wilton	1944	4	3	6
	1969	10.5	b	b

[a]Direct line.

[b]Indicates insufficient number in category.

ers on the basis of tenure and also compared the 1944 median distance with the 1969 median for Deer Creek and Wilton,[1] but no clear trend emerged from either attempt.

Age of Mover

Migration is highly selective with respect to age.[2] Movers to and within our three areas were from six to fourteen years younger on the whole than farmers who had been on the same farmstead tract five years earlier (Table 10). For 1944, newcomers in all townships who bought were older than those who were renting their new place of residence. Wilton's buyers for 1944 were quite a bit older than their counterparts in Deer Creek and Jackson because the acquisition of land for the Joliet Arsenal drove out many owner-operators who had advanced well beyond the usual years of high mobility. Some then purchased new farms in Wilton Township.

---

[1]Jackson is omitted here because it had only one mover for 1969.

[2]Donald J. Bogue, Principles of Demography (New York: John Wiley and Sons, Inc., 1969), p. 763.

TABLE 10

MEDIAN AGE OF FARMERS, BY TOWNSHIP: 1944 AND 1969

Township	Year	Median Age[a] of				
		All Farmers[b]	Farmers Who Were Farming In Same Place Five Years Earlier	All New-comers	Renting New-comers	Buying New-comers
Deer Creek	1944	44	50	38	35	43
	1969	46	48	36	c	c
Jackson	1944	48	52	38	37	38
	1969	51	55	c	c	c
Wilton	1944	44	48	42	40	51
	1969	47	52	43	c	c

[a]For information on how ages were determined see Chapter V.

[b]Includes also those farmers living outside the township with some of their land inside.

[c]Indicates insufficient number in category.

Why Farmers Moved

So many factors, besides a desire for more and better land, have encouraged families to change farms that we found it impossible to decide whether the movers really improved themselves agriculturally. Some were forced to move. Some left a farm because it no longer suited their farming style or the farmstead their life style. One Deer Creek farmer and his wife maintained the only reason they left his mother's farm was to get away from the terrible fleas which inhabited the barnyard. Others had no special complaints about where they were farming before but believed it was time for a change or time to buy a farm of their own. In this section of Chapter IV we are going to look in depth at several of the factors that set farm families in motion during the last thirty years.

The Joliet Arsenal

The first inkling came in mid-August, 1940, when the Joliet Herald-News reported vigorous efforts were being made by the local Association of Commerce to obtain one of the federal government's new munitions factories.[1] In September word finally arrived; a ten million dollar powder plant was to be constructed in southern Will County.[2] Nobody knew exactly where it would be built, but

> amateur observers believed the worked-over areas of the strip mine fields [near Wilmington] would be ideal for a plant manufacturing explosives. The hills and valleys of the waste lands . . . would provide natural buffers to diminish the damage to any accidents in the powder house.[3]

As it turned out, the government was not interested in spending money and wasting precious time trying to reclaim the devastated landscape west of Wilmington. Instead, it planned to condemn sixty-four sections of first-rate farmland between that city and Joliet.[4] These forty thousand acres would provide the space needed for the previously announced TNT works and for its companion, a fourteen million dollar facility to convert the explosives into bombs, shells, and the like.[5]

Meanwhile, as Joliet's businessmen awaited the cleansing of welfare rolls and the influx of millions of federal dollars,[6] hundreds of affected farmers were thinking about other matters. Of little comfort to the men being displaced was the promise by the government to give them top priority, along with veterans, in filling crews to build and run the plants.[7] A feeble resistance movement sprang up, and crowds of angry farm residents gathered at a mass meeting in Wilmington and in smaller conclaves elsewhere to ask questions and formulate policy.[8] To apprise the general public of their displeasure, advertisements were bought in the local newspapers.[9] To apprise Washington, the farmers

---

[1] "Joliet Acts to Obtain Federal Defense Plant," Joliet Herald-News, August 15, 1940, p. 1. Cited hereafter as Herald-News.

[2] "$10, 863, 000 Plant for Wilmington," Herald-News, September 17, 1940, p. 1.

[3] Ibid.

[4] "U.S. Buying 40, 760 Acres," Herald-News, September 22, 1940, pp. 1-2.

[5] "Arms Plant for County," Herald-News, September 24, 1940, p. 1.

[6] "Commerce Body is Grateful," Herald-News, September 25, 1940, pp. 1-2.

[7] "Munitions Plant to Hire Men Within Next Two Weeks," Herald-News, October 9, 1940, p. 3.

[8] "U.S. Plant to Employ 5, 000," Herald-News, September 25, 1940, p. 1.

[9] See: "A Statement to the Public from the Farmers in the Proposed Munitions Plant Site in the Elwood-Wilmington Area," Herald-News, October 13, 1940, p. 17.

assessed themselves three cents per acre to defray the expenses of their emissary.[1] But in the end, all efforts to dissuade the War Department from taking the land it wanted came to naught.

The aroused farmers did manage, however, to focus attention on the plight of the tenants, a group whose rights had been overlooked during the early stages of the tempest. Although little could be done to ease the anguish of the 62-year-old tenant who cried as he told his landlord "old roots do not thrive when transplanted,"[2] financial aid was needed to tide over those tenants being dispossessed far in advance of the customary March moving day. Before a landlord could sell out, he had to produce an agreement signed by his tenant indicating how much compensation the tenant was to recieve.[3] Agricultural economist Case from the University of Illinois was called upon to help tenant and landlord decide what was due the tenant for years left on the lease, improvements he had made, moving costs, and so forth.[4] Running up to $1, 000 per farm, this compensation money was included in the sale price paid the landlord so he could settle up with his tenant.

Clutching their government checks, arsenal farmers fanned out across northeastern Illinois seeking a place for 1941. Those who agreed to move early received less in the way of monetary recompense but did find it easier to get a new farm near the old one. "I guess we were about the first to hear the bad news," remarked a farmer now living west of Manhattan, "and we really hustled to find this farm." The few tracts in the vicinity still owned by insurance companies from the Depression went quickly as did other hard-to-sell parcels owned by individuals. Land values in Will County, especially in nearby townships like Wilton, increased to as much as double their early 1940 level. The chance for a premium price brought onto the market land such as that in estates which could not be settled earlier at "distress market prices."[5] Rental farms around the plant were sold out from under their 1940 tenants or the rents increased. A man

---

[1]"Farmers Want More Money for Land," Herald-News, September 30, 1940, p. 1.

[2]"Gives Land Option for $375, 000," Herald-News, October 1, 1940, p. 1.

[3]"U.S. Fixes Wage Scales on all Construction," Herald-News, October 13, 1940, pp. 1-2.

[4]"Value of Farms Increased by U.S. Purchases," Herald-News, October 20, 1940, p. 26.

[5]Ibid.

who had rented a 280-acre farm in Wilton near the Peotone line on a three-year lease beginning in 1938 recalled how his landlord demanded a larger share of the crop and offered to pay fewer expenses in the 1941-43 lease. The tenant refused to sign. After looking for another place, he gave up and went to work at the munitions plant. "Farmers were wild to get a farm in the winter of 1940-41," he said, "but we knew the place we were renting would not pay if the terms were altered in favor of the lord." Besides the change in terms, there was another problem for him to face--lack of help. Good hired men were quitting the farmers to take better paying industrial jobs. So even though his wife optimistically kept a lot of their agricultural paraphernalia, he never went back to the land after 1940.

Where did the arsenal refugees resettle? We know that five families moved into Wilton Township; and from conversations with local people who have kept track of old friends, it seems that perhaps half of all refugees who continued to farm managed eventually to take up new farms within twenty-five miles of the plant. A few used this opportunity to acquire land in distant localities. Offered, for example, in the October 20, 1940, issue of the _Herald-News_ were farms for sale by the Scully Estate south of Kansas City, farms in the Red River Valley of North Dakota, farms in north-central Indiana and in Ohio.[1] By contrast, the most remote offering in the comparable Sunday issue of the same newspaper a year earlier was just twelve miles from Joliet.[2] Relatives and friends banded together on several occasions to move _en masse_ to new areas. By setting up again in proximity to one another, these families softened somewhat the feeling of isolation engendered by the move.

How now do those who endured the arsenal takeover feel? A bit of the bitterness still persists, to be sure, but for the most part the dislocation is no longer so wholeheartedly condemned as it was during those confused weeks late in 1940.

> We were very upset then because, even though only tenants, we felt like the place would have been ours as long as we wanted to rent it. But in December, 1941, when they hit Pearl Harbor we were glad the plant was there.
>
> Our new home was only a year old when they bought us out. Dad was especially hurt because his land was not used for anything and grew up in weeds. He would drive over there in the summer just to see the old place. It was a bitter experience for all of us, but we made a profit on the land transactions.

---

[1] Classified section, _Herald-News_, October 20, 1940, p. 39.

[2] Classified section, _Herald-News_, October 22, 1939, p. 27.

> We felt bad at the time, but it turned out pretty good. Most of us bought more land than we had in the arsenal and then got to rent it back from the government for practically nothing.

After the War

The immediate postwar period in all three of our areas was marked by a flurry of turnovers when farm-hungry young men returned in search of what a movie produced for the Sinclair Refining Company called Heaven with a Fence Around It.[1] Tenants who had been able to hang onto places during the war found themselves pushed off by enthusiastic veterans. According to one former Deer Creek operator deposed in this way:

> A landlord in those days could get any commitment from a renter. If you were already doing certain extra things [i.e., not specified in the lease], somebody else would agree to do more. The vet who took the farm from me promised to put in a bathroom, rearrange the fences, and generally make the lady's homeplace shine. Besides, they went to the same church.

In addition to their youth and perhaps special ties with landlords, veterans also had the opportunity to participate in the government-sponsored on-farm training programs being set up just for them.[2] While other servicemen were getting a chance to obtain a college degree, willing rural returnees were given instruction in all facets of the agriculture common to their area.

Creditor Ownership of Farmland

Although creditors never controlled "nine out of ten farms in Putnam County," as one former Jackson farmer believes, they did own more than three thousand acres in his township in 1939. Their attitudes toward rental and sale of farm property had for several years a significant impact on interfarm migration practices across certain sections of the Midwest. Farmers had been beguiled by relatively good times after World War I into investing when possible in land of their own.[3] Too often, as it turned out, loan appraisals tended to overvalue poor land and undervalue good land. Murray wrote:

---

[1]Advertisement for movie, Republican, December 26, 1945, p. 2.

[2]See: George William Wiegers, Jr., "Some Outcomes of Participation in Institutional On-Farm Training in Missouri" (unpublished Ph. D. dissertation, University of Missouri, 1949).

[3]A retired Jackson farmer recalled how Unionville's several bankers practically begged farmers to borrow money and buy land.

> When prices dropped in the early thirties, the mortgages on the best land survived the test while the poor land did not produce enough to pay the required interest and eventually came into the hands of the mortgage lenders. It is unfortunate that the poor land was over-valued, because this is the type of land which suffers most from erosion and hard treatment before and during a foreclosure.[1]

Distress was most noticeable and widespread in three areas of the country: (1) the eastern Cotton Belt; (2) the northern Great Plains; and (3) "a portion of the Corn Belt, especially northern and southern Iowa, western Minnesota, and northern Missouri. . . ."[2]

The peak period for creditor ownership of midwestern farms came in the late 1930's. One insurance company's portfolio of farms in 1939 included 400 in Illinois, 284 in Indiana, and 119 in Missouri. By 1942, those figures were 30, 6, and 1, respectively.[3] The Equitable Life Assurance Society, a major farm mortgagee prior to the Depression, topped out with 6,065 units in its portfolio in 1938.[4] The Metropolitan Life vice-president who in 1939 was dubbed "America's Biggest Farmer" saw his two million acres dwindle to only a couple hundred thousand by late 1942.[5] For Jackson the peak may well have occurred in 1939; for our other two townships it was a bit earlier. Roughly one-seventh of Jackson land belonged in 1939 either to life insurance firms or to a variety of other creditors (Table 11). Mainly nonlocal, these other creditors included real estate firms, bankers, and shrewd investors who had bought greatly discounted mortgage notes from failing banks a few years before.[6] Some of the Jackson tracts still remained on creditor rolls in 1944,[7] but the four in Wilton and two in Deer Creek were by then back in the hands of farmers. Of the four

---

[1] Murray, Farm Appraisal and Valuation, p. 483.

[2] Ibid., p. 484.

[3] "Farms Unfrozen," Business Week, August 8, 1942, p. 44.

[4] F. J. Skovold, "Farm Loans and Farm Management by the Equitable Life Assurance Society of the United States," Agricultural History 30 (July, 1956): 114.

[5] Ladd Haystead, "Biggest Farmer Biggest No More," Fortune, December 1, 1943, p. 74.

[6] Interview with Henry Gardner, former local field agent for Connecticut General Life plus numerous private creditors, Unionville, August, 1969. Interview with Elias Shuey, former local field agent for Union Central Life, Unionville, October, 1969.

[7] In 1969, a handful of small, inaccessible, undesirable timber tracts were still in creditor hands.

TABLE 11

LAND IN THE HANDS OF INSURANCE COMPANIES AND OTHER CREDITORS, BY TOWNSHIP: 1939

Township	Number of Tracts	Acreage	Percent of Total Township Acreage
Deer Creek	2	220	1
Jackson	29	3,180	14
Wilton	4	705	3

in Wilton, two had been bought by arsenal refugees and a third by a family with money inherited from a father who sold out to make way for the plant.

These absentee owners encouraged both voluntary and involuntary farmer turnover. First, people moved out voluntarily when they found the land unproductive or because of dissatisfaction with improvements (or the lack of them). "Loan companies," said a Jackson widow, "were okay to rent from but they wouldn't fix anything. Fences and buildings went to pot." Second, farms were sold whenever a suitable buyer could be found. There existed the belief that a farm was "half sold when rented to a good tenant."[1] Thus the field agents representing the insurance companies and private noteholders tried to size up a renter in the first couple of years on the place. If he seemed promising, they might offer him the chance to buy. When a man declined the purchase alternative, agents had a tendency to invoke the three-year or five-year clause in his lease and force him to move. In this way they exposed their farms to a series of prospective buyers.

Getting Out of the Mud

According to a Bureau of the Census map dated April 1, 1950, slightly less than one-third of American farmsteads were located on dirt roads.[2] Will and Cass counties were classified in the "Under 20 Percent" bracket, but in Putnam County and neighboring Sullivan County to the south somewhere between 60 and 79 percent of the farms found their rutted road dusty in the summer and a

---

[1]J. M. Huston, "How Can Delinquent Loans and Foreclosed Properties Best Be Serviced and Handled," Journal of Farm Economics 22 (February, 1940: 282.

[2]The map is reproduced in: U. S., Department of Agriculture, Agricultural Research Service, Current Developments in the Farm Real Estate Market (Washington: Government Printing Office, October, 1958), p. 25.

muddy morass whenever it rained or thawed. Not surprisingly, the desire to escape the dust and especially the mud prompted a fair share of the moves involving Jackson Township farm families during the 1940's. Here is a description of what they were fleeing:

> Most everywhere else in the U.S., mud is a sometime thing, puny, thin and undistinguished. But in the state of Missouri, mud in the springtime is a lusty manifestation of nature--a phenomenon in its own way as deserving of celebration as the falls of Niagara. Missouri mud is rich, muscular and ubiquitous. . . . Upstate Missouri is where the mud lies thickest in the spring. North of Kansas City in the west and St. Louis in the east the land between the Mississippi and Missouri Rivers is seemingly bottomless topsoil.[1]

While mud would restrict a family's shopping trips and stymie the shipment or delivery of farm commodities, it probably had its most telling effect on high school students whose buses ran only on the all-weather roads (of which Jackson had practically none prior to the Korean War). Thus when a bright rural Putnam County teenager finished the eighth grade, his parents had a choice to make. They could either pay to board him out in town (or somewhere convenient to the bus route) as had long been the custom or they could leave the mud behind altogether. Below are comments from three farmers who scrambled out to live on U.S. 136 during the first half of our study period:

> Through the immediate prewar years we were farming south of Lucerne in Medicine Township on 160 rented acres. When the kids got ready for high school, we bought here on the highway [in northern Jackson].
>
> We bought the eighty on 136 and farmed it for awhile from the old place on West Locust. By 1944, however, we had moved up to the highway so that our daughter could get a high school education.
>
> The gravel kept creeping closer and closer to us on the road that led to the big farm east of East Locust, but it was still a half mile away. So we gave it up and bought here on 136 to give the kids easier access to the high school in Unionville.

Landlord-Tenant Disagreements

Most rental arrangements end on a fairly amicable basis, but a few leases are abruptly terminated leaving each party furious with the other. During the course of this research several such incidents came to the author's attention. As told by the tenants, here are the details of three ousters:

> We moved into Deer Creek from north of Logansport to rent the . . . farm at Deacon. Everything went well for a couple of years until the landlord and my wife got into it. Even though the house had no running water, he

---

[1] "Missouri Mud," Colliers, May 5, 1951, p. 28.

> announced plans to install up-to-date facilities, including waterlines, out in the cattle lot. My wife said there would be none of that until her home was modernized. He booted us.
>
> For several years we had rented south of Unionville, in Wilson Township, sharing the government crop diversion payment. Then our elderly landlord got the idea of taking it all himself without reducing the rent accordingly. We balked, and he gave us notice to move out.
>
> I had been farming north of Wilton Center on the Manhattan line during the thirties. One day in November I came back from hauling manure to find the landlord waiting. He informed me I had too many kids for the house and would either have to pay extra rent for it or leave. We scraped up a thousand for the down payment and bought this eighty.

One likable Deer Creek landlord enjoys a certain amount of local notoriety because of the number of tenants he has wooed and then shooed. The standard tenure period was two years. During the first year most renters were trying to figure him out and, as a neighbor put it, "during the second they were on their way out." On his three small rental farms the landlord maintained excellent buildings and fences which made an immediate impression on the prospective lessee. Obligations of the tenant were, as they should have been, spelled out precisely in the lease. Not spelled out, unfortunately, was the landlord's propensity to demand that livestock come before everything else, including crops and family. Though a veteran farmer himself, he knew, according to a former tenant, "a lot about feeding stock but practically nothing about field work. Appearance of the field meant more to him than production." At least one of his tenants grew so tired of the unreasonable demands and unannounced visits that the sheriff was called upon to evict the landlord from his own property.

Moving as a Way of Life

The wives of contemporary corporate nomads, Packard tells us, "assume they will soon be moving on again, and view the prospect with varying degrees of equanimity."[1] Few will "burst out laughing when someone says IBM means 'I've Been Moved.'"[2] They have heard it too many times before. Not so long ago many midwestern farm wives found themselves in much the same predicament. Moving for the wife of a tenant in Deer Creek or Jackson or Wilton during the 1930's, 1940's, and into the 1950's was an accepted part of her alliance with the land. And, likely as not, her husband had resigned himself to it, too.

---

[1] Vance Packard, A Nation of Strangers (New York: David McKay Company, Inc., 1972), p. 29.

[2] Ibid., p. 18.

Luckily there was little or no stigma associated with frequent interfarm moves.[1] "We were married in 1929," said a widow now living in Unionville, "and for twenty years we went from one farm to another. It was expected of poor people like us." "Most movers in those days were young," another Unionville widow noted, "and no one gave it a second thought." According to a Jackson farmer who had been on a dozen or more farms during his adult life, "There were some where you just couldn't make any money. We would stay a year, maybe two, and leave." Moving was something that had to be endured if a family ever hoped to get ahead and buy a place of its own. When a tenant did settle down on one farm for a considerable length of time folks had a tendency to forget he was just using it. In other words, tenure longevity was to ownership as mobility was to renting.

Finding the Next Farm

The midwestern farmer has depended on a number of sources for information about farm availability. Like most marketplace participants, however, he has suffered in his pursuit of a better situation from incomplete knowledge of what was for rent or for sale at that moment. Likewise, those with land to let or sell have been faced with the same dilemma. To alleviate some of this uncertainty, Timmons suggested a landlord-tenant exchange file for each Iowa county.[2] Farms were to be described by the owner for potential tenants to consider. Farmers, to aid landlords in their search, would jot down qualifications, references, and needs. Other than a pair of pilot projects, nothing ever came of this intriguing plan. It is not surprising. First, there would have been a modest participation cost. But even more important must have been the usual reluctance of farmers to admit they were looking and of owners to admit they were thinking of retiring as farmers or of changing tenants.

Many transactions involving sale or rental of farmland occur without the knowledge of anyone except the parties directly involved and, in the case of a sale, their attorneys. A farmer from near Wilton was waiting to unload grain at the elevator one fall when he struck up a conversation with a stranger waiting to do the same. As it turned out the second man had just bought a large tract in

---

[1] One elderly Jacksonian, recalling a bright side of the turnover situation, noted how they were always happy to have new neighbors.

[2] John F. Timmons, Improving Farm Rental Arrangements in Iowa, Iowa Agricultural Experiment Station Research Bulletin 393 (January, 1953), pp. 83-84.

Wilton and wanted someone to farm most of it. By the time the two reached the scales, a tentative decision had been made for the Wilton-area farmer to rent 310 acres. A Deer Creek landlord frequented a cafe on Indiana 29. There over coffee in 1956 he met for the first time a farmer from Clinton County.[1] When they parted, the farmer had agreed to leave his rented farm and bring the family to live and farm in Deer Creek Township. Both of these encounters represented windfalls since neither farmer had been actively searching for another tract of land. In his Philadelphia study of family movement, Rossi found the windfall an extremely effective information source.[2]

Rossi mentioned the neighborhood storekeeper and the friend with a vacancy in her building as good urban information intermediaries.[3] Our farmers found business acquaintances, relatives, and friends helpful, too. A Deer Creek tenant heard of his first Deer Creek farm through a friend while living north of Logansport. The friend learned of it from the Cass County sheriff who had just evicted the previous renter. Another Deer Creek transaction began to unfold when a family from Miami County having their automobile gassed at a Galveston station casually mentioned their desire to buy a place. A widow, whose eighty they would later acquire, had left word with the station attendant that she might be willing to think about selling. Bankers, implement dealers, elevator operators, and others to whom farm people come for services were also useful go-betweens. At the same time, men who brought their products and services to the farmstead acted as information carriers. A farm family then living at the east end of Putnam County learned of a place in Jackson through the trucker who was going to haul cattle to St. Louis for them. Other carriers of information included mailmen, feed salesmen, petroleum distributors, and veterinarians.[4]

Few farms, compared to the number changing hands, were advertised in the local newspapers of our areas.[5] Individuals prefer personal contact; and besides, ads cost money. Seldom are "For Rent" and "For Sale" notices posted.

---

[1]Clinton County is two counties south of Cass County.

[2]Peter H. Rossi, _Why Families Move: A Study in the Social Psychology of Urban Residential Mobility_ (Glencoe, Illinois: The Free Press, 1955), p. 160.

[3]_Ibid_.

[4]Even the present author was quizzed occasionally by farmers about land availability elsewhere in the study areas.

[5]Selected 1939 and 1969 issues of both the _Republican_ and _Vedette_ were checked for farm advertisements.

Realtors employ the classifieds on a more or less regular basis, but even they rarely post availability signs to alert the passing public as is so common in urban and suburban situations. Farms are sometimes listed in the classified section of farmer-oriented periodicals. This avenue is especially attractive if someone wants to contact a buyer or renter for a very expensive or highly specialized farm. A Kankakee farm manager noted that his firm would occasionally advertise in the Prairie Farmer if, through the usual local channels, it was unable to find the right man for a farm being managed.[1] Another steady patron of the Prairie Farmer classifieds has been the Deer Creek landlord mentioned earlier who had so much trouble keeping tenants. Neighboring farmers already knew his eccentricities too well. Also in the farm magazines is an occasional solicitation for a farm. The following appeal appeared as we were completing the Deer Creek phase of the field work:

> WANTED TO RENT: Sober Christian couple with 3 grade-school-age boys want to rent low-pressure type of farm (prefer central Indiana) where we can have some livestock, hay and pasture. Not interested in raising or feeding 5000 hogs or cattle or 50, 000 chickens; tired of 600-800 acre corn and beans farms; practical farm family with good health, good equipment, 25 years experience. . . .[2]

Hopefully, they found it.

When Moving Day Arrived

Corn Belt leases customarily expire at the end of February making March 1 the day when families with unrenewed leases must vacate.[3] Looking across his Iowa farm at a neighbor loading a truck similar to the one shown in Plate I, Wolf lamented a tenure system which that day was forcing perhaps fifteen thousand other Iowa renters to do likewise. Some would have to unload in town, and many who did succeed in finding another farm would have to take a place that was unsuitable because of size or facilities to their wishes and capa-

---

[1]Interview with Gary Kendle, Farm Management Specialist with City National Bank of Kankakee, Kankakee, Illinois, January, 1970.

[2]Classified section, Prairie Farmer, August 1, 1970, p. 59.

[3]The lease expiration date is a function of climate, type of farming, tradition, and perhaps other factors. New Year's Day was once sharecropper moving day in the South where work on the cotton land had to begin very early in the year (John I. Clarke, Population Geography [2nd ed.; Oxford: Pergamon Press, 1972], p. 136). It might be both instructive and interesting to map the United States county-by-county according to the prevailing dates of lease expiration.

Pl. I.--A tenant farmer prepares to move off the place he has been operating near Lafayette, Indiana. (Photo courtesy of the United States Department of Agriculture.)

bilities.[1] Exactly three decades later, Collins was recalling March 1 in these words:

> Moving Day! Cornbelt, USA, March 1st, the traditional day of new homes, new contracts, new friends, new hopes. Moving Day; the sad day of adult memories, successes, loves, miseries and failures.[2]

Some moving-day essayists advised farm folk about ways to ease the burdens of that difficult day. Brown offered suggestions to the homemaker on everything from how to move the kittens (in a gunny sack) to what foods to fix for the first meal in the new surroundings (baked beans, creamed chicken, potato salad, and coffee). Laying hens could be fooled into maintaining their schedule if "cooped and moved after dark, and uncooped before daylight, to find their familiar waterers and feeders waiting."[3] Wallace's Farmer and Iowa Homestead urged the prospective mover to delay his farrowing until late March and thereby give the sows a chance to grow accustomed to their new home. It was a good idea, the writer went on, to start moving well before March 1 if the tenant at the next place would cooperate. Perhaps he would even allow the incoming farmer to do some fall plowing and fence mending.[4]

Although they commented on individual moves, the country correspondents covering Wilton and Deer Creek never recognized moving day per se as Jackson correspondents regularly did down through the years. Here are a few examples from the Republican.

March 8, 1939	"This end of Route No. 5 had no mail several days last week on account of snow. The drifting snow of February 28 came just in time to delay folks moving March 1."
March 11, 1942	"There is some moving to various places this spring."
March 10, 1948	"Well this is March 1st and the ones that have to move are busy moving."
March 13, 1963	"Several folks were moving the past week to their new locations for the coming year."

Preparations for impending moves caught their attention, too:

---

[1] C. E. Wolf, "March 1st (What a Mere Date Can Mean to a Tenant Family)," Commonweal 31 (March 1, 1940): 404-5.

[2] Eddie Collins, "Good Morning Feeders and Hi Mom!" Republican, February 25, 1970, p. 8. Collins was a syndicated columnist.

[3] Mabel Nair Brown, "Put System into Moving Day," Farm Journal, February, 1947, p. 110.

[4] "If You're Moving Next Spring," Wallace's Farmer and Iowa Homestead, October 18, 1952, p. 38.

February 18, 1942 "Mrs. Eva . . . and Mrs. Anna . . . are helping Mrs. Orla . . . clean the house where she intends to move [with her husband];"

as did arrangements made soon afterward:

March 8, 1939 "Misses Elizabeth and Pauline . . . started school at Central City, Wednesday."[1]

To Miss Elizabeth, Miss Pauline, and thousands of other children, a March 1 move meant a midterm interruption of their education. To the Putnam County families who had moving in mind for 1939, it meant waiting out the effects of a late winter snowstorm. But perhaps the most memorable moving-day misery for veteran Jacksonians was the very thing some were trying to flee--the mud. From the standpoint of dirt road usability, no worse time of the year could have been chosen for interfarm migration. Unionville's normal daily temperature during the week of February 26-March 4 stands right at 32°F.[2] This, of course, meant a solidly frozen road at 5 a.m. was apt to be impassible when the temperatures climbed into the forties by midafternoon. For good reason, many moves were made by moonlight.

Sometimes Jacksonians were delayed not by mud in front of their own farmstead but rather by the mud or some other problem affecting a family ahead of them in that particular movement chain.[3] A new baby at the wrong time of the year might force postponement of a half dozen migrations because the parents were unprepared to vacate in favor of their replacement. Equipment and stock could be shuffled in and out; but unless people were willing to share quarters temporarily, it might be a couple of weeks before everybody got settled down for the spring work. Figure 8 depicts a Jackson chain which, according to one of the links (Farmer E), just happened to go smoothly. Farmer A left a small farm and went over to his father's 130-acre tract at the western edge of Jackson. The man he replaced, Farmer B, trekked across West Locust to a comparable tract which he had inherited earlier. His brother, Farmer C, moved down the lane to an eighty that he bought from an insurance company. This transaction forced its renter, Farmer D, to look for another farm. He

---

[1]The Central City School was located just west of our Missouri study area on U.S. 136. It now shelters bales of hay instead of scholars.

[2]Stephen Sargent Visher, Climatic Atlas of the United States (Cambridge: Harvard University Press, 1954), Pl. 52.

[3]Movement chains are not confined to the country. See: John B. Lansing, Charles Wade Clifton, and James N. Morgan, New Homes and Poor People: A Study of Chains of Moves (Ann Arbor, Michigan: Institute for Social Research, 1969).

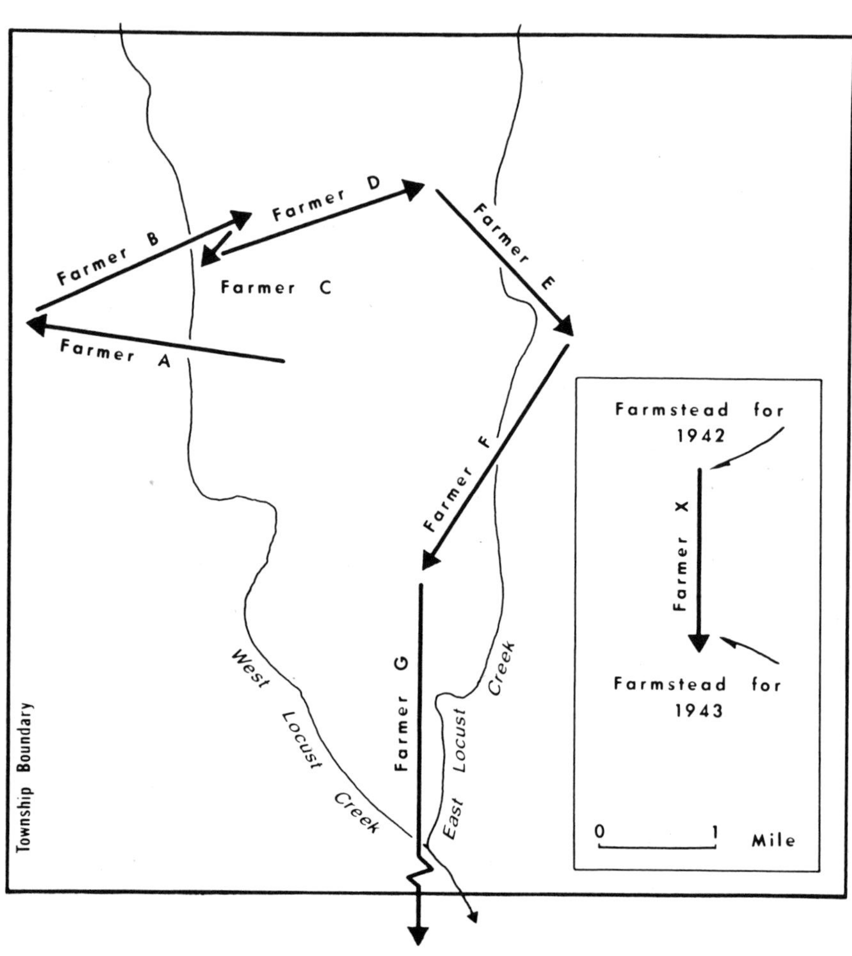

Fig. 8.--An Interfarm Migration Chain, Jackson Township: For 1943

found it on the bluff overlooking East Locust. Farmer E had already decided to vacate that place for a much larger bluff and bottom farm across the creek. Farmer F, who had been there in 1942, rented an even bigger parcel farther downstream. He replaced Farmer G who left the township for a tract many miles to the south. The fate of Farmer H and those ahead of him in the chain is unknown.

A Final Comment

Why do farmers move less today than they did yesterday? There are, of course, fewer potential movers because of the reduction in farmer numbers; but as we have seen earlier in this chapter, a decided decline in the proportion of all farmers who move between focus years has occurred since the early 1940's. First, the relinquishment of land by creditors has brought it into the hands of more affable landlords with whom their tenants can form lasting partnerships. Better still, the farmers have purchased or inherited the land themselves. Second, construction of all-weather roads (and extension of school bus routes) in Jackson has removed another reason for relocation. Third, many of the small rental farms on which young couples could start out have now been added by other farmers to their holdings. In this way, a potentially mobile group has been denied the chance to farm. Fourth, the difficulties and complexities of modern agriculture have weeded out veteran farmers who were just good enough to get a farm but not good enough to keep it more than a year or so. Finally, men who were once highly mobile found they were gaining little by moving around so much. Consequently, decisions were made to buy a farmstead tract or to stay on a good rental farm. In the next three chapters of this report we will pursue the matter of farm expansion from a secure farmstead tract.

CHAPTER V

NONFARMSTEAD LAND: FIRST CHOICE OF THE CONTEMPORARY FARMER

When in need of additional acres, the midwestern farmer typically reacts by shopping first in the land markets that can bring him into contact with landowners who have nonfarmstead tracts to lease out or sell. With declining frequency, as we learned in Chapter IV, does the established farm operator seriously consider migrating from one farmstead to another to obtain more or better land. The farmstead tract encompasses all the land surrounding the base of operations which is owned by the person owning the actual farmstead. That person may be the farmer himself or a landlord. Any other land in the farm, either lying adjacent to the farmstead tract and owned by someone other than the farmstead owner or lying away from the farmstead tract, no matter who the owner might be, is considered nonfarmstead land. According to this breakdown, there are only two categories of farmland--farmstead and nonfarmstead. In the last half of this chapter, however, we will separate for special treatment those nonfarmstead tracts that are not contiguous with the farmstead acreage.

Farmstead acres in 1939 outnumbered nonfarmstead acres roughly three-to-one in our midwestern study townships (Table 12). The great majority of 1939 farmers had all their land in one block, and quite often they farmed for only one owner (Table 13). By 1969, nonfarmstead acreage surpassed farmstead acreage in Deer Creek and Wilton while approaching equality with it in Jackson. This increase was accompanied by a sharp decrease in the number of farmers with one-owner, compact units.

Acquisition of Nonfarmstead Land

The reader will recall that farmers can obtain land in three ways: inheritance, purchase, and rental. For nonfarmstead land, the rental alternative is by far the most important. Since the other two methods of acquisition are relatively insignificant, we can quickly dispense with them.

TABLE 12

FARMSTEAD AND NONFARMSTEAD ACREAGE AS A PERCENT OF TOTAL TOWNSHIP ACREAGE: 1939 AND 1969[a]

Township	Percent of Total Township Acreage that Belonged to a			
	Farmstead Tract		Nonfarmstead Tract	
	1939	1969	1939	1969
Deer Creek	70	37	30	63
Jackson	73	51	27	49
Wilton	74	39	26	61

[a]Here and elsewhere in this chapter, data are for the study townships themselves plus any operating units with land on both the inside and outside.

TABLE 13

OPERATORS WITH A COMPACT UNIT WHO WERE FARMING FOR JUST ONE OWNER AS A PERCENT OF ALL OPERATORS, BY TOWNSHIP: 1939 AND 1969

Township	Percent of Operators with Compact, One-owner Units	
	1939	1969
Deer Creek	55	20
Jackson	57	24
Wilton	52	17

Inheritance

As a percent of all nonfarmstead acres, inherited acres ranked last in four of the six cases: Deer Creek in 1939 and 1969, Jackson in 1969, and Wilton in 1969 (see Table 14).[1] There were in Deer Creek during 1939 (the extreme case) only two farmers (out of 175) who had inherited and were farming any nonfarmstead land. One had gotten a thirty-three-acre tract which was once a part

[1]If part interest in a tract was inherited by a farmer or his wife and the balance later purchased, we consider the entire tract to have been inherited.

TABLE 14

METHOD USED BY FARM OPERATORS TO OBTAIN CONTROL OF FARMSTEAD AND NONFARMSTEAD LAND, BY TOWNSHIP: 1939 AND 1969

Township	Year	Percent of Farmstead Land		
		Rented	Bought	Inherited
Deer Creek	1939	46[a]	30	25
	1969	45	27	28
Jackson	1939	32	32	36
	1969	10	62	27
Wilton	1939	69	12	19
	1969	67	16	17
		Percent of Nonfarmstead Land		
		Rented	Bought	Inherited
Deer Creek	1939	86	13	1
	1969	83	13	4
Jackson	1939	66	12	21
	1969	56	34	9
Wilton	1939	83	2	15
	1969	84	11	4

[a]Rounding off may cause lines to total more or less than 100 percent.

of his homeplace, and the other had received two tracts totaling seventy acres that lay a couple of miles from the large farm he was renting. These 103 acres represented just 1 percent of Deer Creek's nonfarmstead land in a year when the farmstead acreage inherited by Deer Creek farmers accounted for 25 percent of that category.

Why is nonfarmstead land so seldom inherited by farmers? In the first place, benefactors often have only one tract to bequeath--the family's farmstead tract. Second, if a benefactor has more than one tract to pass down, he or she (usually a parent) would rather give their farmstead to a son or daughter who is planning to continue farming than to other children who would not appreciate the buildings as much as the farmer. Land remaining after the farmstead tract is assigned can be given to nonfarming progeny who are thereby saved the trouble of finding suitable occupants for the buildings. A piece of this bare farmland, by the way, will bring in as much or more rent per acre in many areas than will similar land having a good set of buildings. In summary, if a farming heir inherits any land, it is apt to be a farmstead tract although there are, of course, exceptions. Once he has this as a secure base, he can more confidently venture out and obtain other land by rental or purchase.

Purchase

When a farm family has saved enough money to buy land, it is traditional to invest in a tract having a set of buildings. With the land the family also gets a place to live, to protect and repair machinery, and to keep livestock or do whatever it chooses without fear that it might displease a landlord and end up out of agriculture completely. If already established on its own farmstead, the chances are good that a family would pay a premium price for an adjacent parcel, which would augment the farmstead acreage, but less for a tract lying elsewhere. As one young, expanding Wilton farmer living north of Wilton Center who had just bought an adjacent eighty put it, "you may get only one crack in a lifetime at ground next door so you best not have your money or credit tied up someplace else when the opportunity comes along."

Rental

For accumulating nonfarmstead land, the lease is the overwhelming choice of midwestern farmers. It is by design an ephemeral association, and each party knows the other could on short notice dissolve their fragile tie. Because of this very ephemerality, the lease is a perfect device when additional nonfarmstead land is needed for a short or indefinite period of time. With a lease a Jackson cattleman can secure the right to make extra hay in years when his own is likely to be insufficient. The young Deer Creek hog farmer can temporarily rent enough extra land to pay for the new tractor or new baby. A short-term lease on his cousin's land gives the Wilton father a chance to determine whether his sons are really interested in becoming and apt to succeed as farmers before he buys more land and machinery with them in mind. At the other end of the farmer's life, rental land can be easily sloughed off without further responsibility until only the farmstead tract is left.

Share Rental

Most cropland leases still involve some sort of share-the-cost, share-the-product arrangement. The lessee furnishes labor, equipment, and fuel; the lessor provides the land and pays the taxes on it; and both share the cost of fertilizer, seed, pesticides, and perhaps other expenses depending on the agreement. Almost invariably the landlord and renter share equally in crops harvested and in money paid to the farm by the federal government for not planting certain crops. Since any buildings, other than grain storage facilities, on nonfarmstead land rented for crops are superfluous as far as the tenant is concerned, the share landlord collects from him no cash compensation for them. The owner might, however, be able to rent the house out to a family that wants to live in the country badly enough to endure such annoyances as being awakened on Saturday mornings by the growl of a struggling diesel under the bedroom window.

Among agricultural economists, the practice of leasing nonfarmstead land is called field renting, no matter how many actual fields in the usual sense of the word are involved. They are differentiating, of course, between farm rental which includes the buildings, and rental of only the land, or fields. Nonfarmstead (field) rental arrangements between farmers and landowners in Deer Creek and Wilton typically take in the whole tract exclusive of any buildings. The practices of carving up a tract by letting more than one man rent portions of it at the same time or of allowing someone to rent a small field here and there are largely unknown. When a man gets ready to take another job or to retire from farming, he will usually cut back his operation by the tract instead of by the field. Many in these two tillable townships feel that fields are already too small and rows already too short to have landlords further restricting the big machines.[1]

The situation in Putnam County, Missouri, is somewhat different. If Deer Creek and Wilton farmers practice field renting, then Jackson farmers have a predilection for patch renting. Many of the cattlemen in Jackson need only a few acres of corn or soybeans each year to satisfy their grain requirements. Parcels become available to them in the following ways. First, there are a number of older farmers (and some younger ones working in town) who can look after their own livestock but periodically call on a patch renter to plow up a hillside, harvest a crop or two of corn, and then seed it down with a good stand of grass. Second, there are always small, inaccessible patches scattered along the creeks and branches where a sanguine man can plant a few acres of soybeans

---

[1]For a discussion of steps being taken in one area to enlarge fields see: John Fraser Hart, "Field Patterns in Indiana," Geographical Review 58 (July, 1968): 450-71.

after the gumbo finally dries out.[1] The hope is that the beans will go forty bushels to the acre and not down the creek around Independence Day.[2] Patch renting, for obvious reasons, is shunned by those Jackson farmers who are assiduously attempting to expand their operations; but it does meet a real need of their less ambitious neighbors.

Cash Rental

Not all farmers in these three areas rent their nonfarmstead land on a share basis. Cash rent is becoming more important each year. It is, for instance, the only logical way to handle nonfarmstead pasture rentals in Putnam County. Increasing numbers of Deer Creek landlords have begun to request cash rent rather than a share of their tenants' crops. At fault in the opinion of the Deer Creek farmers being forced to pay the $45-$55 per acre rates are

---

[1]"FOR RENT: 16 acres of bottom ground, 2 miles north of Hartford, Mo. . . . " Classified advertisement, Republican, April 2, 1969, p. 6.

[2]A late June or early July flood along the Locusts is not uncommon. While this author was there in 1969, the water came out of the channels just about on schedule (see Plate II).

Pl. II.--View of flooding along West Locust Creek in southwestern Jackson Township early in July, 1969.

(1) local tomato growers, (2) nearby seed companies, and (3) land-hungry young farmers who would do almost anything for more land. Some unhappy cash renters, especially in the southeastern portion of the township, near the tomato cannery in Galveston, blame exorbitant cash rent requests on the $60 per acre that tomato producers gladly pay for land they rent. A landlord hears about the tomato rentals, does a little calculating, and demands similar sums from his renters even though the soil and location of his tracts might be completely unsuitable for tomatoes. Farmers near Young America and south toward Burlington and Camden are forced, in some instances, to compete for land with companies growing hybrid seed.[1] To produce their valuable crops seedsmen, too, will pay $60 per acre for land. Finally, Deer Creek farmers are faced throughout their area with the threat of cash rent because landlords have taken it from expanding young farmers and found it more desirable than share rent. With cash, both the landlord and the eager expander are satisfied. One gets the land on which to test his skills, and the other gets a guaranteed income without having to worry whether or not the untried youngster will ever get the corn planted. The only dissatisfied party is the average farmer who must compete for the same land and please the same landlord.

In the vicinity of Wilton Township, cash rent is still pretty much confined to the land owned by two institutions--a state mental hospital near Manteno in Kankakee County and the Joliet Army Ammunition Plant in Will County. Surrounding the Manteno State Hospital, located seven miles south of Peotone, are 500 acres which the State of Illinois in the late 1960's began leasing to local farmers for about $50 per acre. Wilton farmers found the Manteno land a bit too remote[2] for them to consider, but enough information had seeped back about the rental fees to cause some in our study township to question the sanity of those who did. Although only a small proportion of Wilton farmers have ever rented land offered by the federal government at the Joliet Arsenal, most concede that it does represent an important alternative if one is ready to tolerate the restrictions imposed on the lessee. We will, therefore, examine here in some detail the landleasing program at the JAAP. Our information comes from

---

[1] Corn and soybeans.

[2] The hospital land is remote from Wilton not only because of distance but also because (1) it is in another county; (2) it is in an unfamiliar part of that county; and (3) it lies beyond I-57, Ill. 54, and the Illinois Central Gulf mainline all of which help to create a psychological barrier.

official documents; interviews with government spokesmen; and interviews with Wilton farmers who do, did, would, or would never rent land "over in the plant."[1]

Leasing of Land at the Joliet Arsenal

In 1969, of the 24,000 acres remaining in the hands of the United States government at the Joliet Arsenal, 16,000 were rented to farm operators who had outbid other men for the privilege of planting 1,000 acres of popcorn, making hay from 6,000 acres, and running beef cattle on the remaining 9,000 acres. Altogether there were seventy-nine rental tracts ranging in size from a forty-acre popcorn tract in the ammunition-storage sector just west of Wilton to thousand-acre and larger pasture parcels in the manufacturing zone between Illinois Highway 53 and Interstate 55. Since farmers cannot live on their land in the plant, all seventy-nine tracts fall into our nonfarmstead category. Of the seventy-nine, eleven were leased to men who also were farming land in Wilton. One other tract included the sixty-five acres taken by the government in 1940 from the northwest corner of our township.[2] Before looking further at the contemporary leasing situation, however, perhaps we should return to the early arsenal years and trace the development of agricultural leasing from its inception.

As the United States geared for and began fighting World War II, it accumulated a quantity of good farmland on installations around the country. The Joliet Arsenal, then totaling over 40,000 acres of which a fifth was actually needed for the manufacture and storage of explosives and four-fifths served as a safety shield to protect surrounding areas, was one of the largest of these installations. Originally, the War Department planned to seed down the open space at Joliet with grass and to keep a team of tractors mowing throughout the growing season. To this end, thousands of acres were put into grass in 1941 and 1942. Regrettably, thousands of these acres grew up in weeds which went to seed and spread beyond the perimeter fence to the nearby fields of local farmers who were already unhappy with the government for deciding to build the arsenal there in the first place. Furthermore, the dry weeds increased the likelihood of fire in an environment where any open lights were looked upon with disfavor. Finally, neither grass nor weeds did much for the food production efforts of an

---

[1]Sources of commonly known information about the arsenal will not be footnoted.

[2]Information on lessees is from: Joliet Army Ammunition Plant, Office of the Land Manager, <u>Mailing List and Telephone Numbers of Agricultural Lessees</u>, Elwood, Illinois, April, 1969.

embattled nation. Therefore, early in 1943, the maintenance crew made plans to plow up and farm the land they had been trying to keep mowed in the previous two years. Projected expenses for equipment and the lack of help soon convinced them, however, that other means had to be found to get the job done. With little choice left, the War Department turned to the farmers for help.[1]

For the next fifteen years, farmers were allowed to plant almost anything they wanted so long as they followed a few simple rules. At the outset, lessees paid only a couple dollars per acre cash rent and agreed to control the weeds. Since there were restrictions on the height of crops in strategic locations, corn, which might have served to conceal a force of saboteurs, sometimes gave way during the war to shorter crops like soybeans, wheat, and oats. During those early years a harassed land management office had to contend with all sorts of problems not the least of which was the crafty Irishman who had been moved from a farm in the eastern part of the arsenal to another just across the fence in Wilton. From there, along with his two sons, he rented a great deal of land in the conventional manner, but, according to local legend, also farmed some that other men had rented and even found patches to farm that the officials did not know existed or that other farmers feared to farm because of proximity to explosives.

Land management officials tried early on to secure permission for a share-lease system, but their superiors refused because of the additional supervision that would have been needed. Reflecting on this rejection, one of the men who had fought for share rent told the author that it was probably just as well they had failed. The cash lease never proved as alien to farmers as had been expected.[2] One Wilton lessee confided that he was initially dubious about paying for land that might yield nothing but now he prefers it, perhaps because this means he has at least some land which is not owned by a snoopy landlady. Why did the cash lease fail to spread from the arsenal to privately held land after more people became familiar with it? For one thing, only a minority of Wilton farmers have leased JAAP land; and they, in order to discourage competition for the land at subsequent biddings, have neglected to praise the system enthusiastically. Beyond that, cash leasing at the plant is confined to a special area

---

[1]Interview with Harlow Nicholson, former foreman of land maintenance crew at the Joliet Arsenal, February, 1970.

[2]Interview with John L. Kirkton, Chief of the Appraisal Branch, Real Estate Division, Chicago District, U.S. Army Corps of Engineers, June, 1969.

and situation that makes comparisons by outside landowners more difficult than in Deer Creek.

The years when arsenal lessees really "cleaned up," according to those who bid unsuccessfully or were too young to bid, lasted until Secretary of Agriculture Benson issued a decree in May, 1956, forbidding future use of government land to produce price-supported crops. Between the end of World War II and the Benson ultimatum, the leasing program had moved from its chaotic infancy into a period during the late 1940's and early 1950's when the land managers finally found who was farming what and were able to see that land improvement measures specified in the leases were carried out. Lessees were required to follow a crop rotation program prescribed by the land management office for the five-year lease period and to fertilize according to the needs of each particular tract. They had to adhere to a whole array of minor nuisance regulations concerning safety and security. Despite these restrictions, sometimes arbitrarily enforced, farmers competed vigorously for arsenal land.

After the 1956 decree, corn, soybeans, and wheat gradually disappeared as grain leases expired. By the February, 1960, deadline most grain tracts had been converted to hay or pasture tracts. To stimulate interest in livestock, the land management office consolidated a number of tracts to make larger units and to take advantage of the installation's security fences. For the first time the federal government authorized ten-year leases of pastureland because of a fear that some of the land might go begging for a livestock man to use it. University of Illinois animal husbandry specialists conducted meetings for the crop-oriented local farmers hoping to convince them that running beef cattle on government land might be a profitable sideline. This excerpt from a Corps of Engineers memo nicely sums up the government's predicament in the late 1950's:

> Prior to the [Benson] Memorandum, the leasing program was the lessor's market with lessees growing the high income cash crops, realizing a quick return and demanding more land as a result of the increased productivity of the individual operator. Now, however, with the increased amount of pasture in a concentrated area with the lessee assuming the great expense of establishing pasture, adequately stocking it, waiting several years for a return, performing the required maintenance, and in the interest of good pasture management, it is in the interest of the Government to make the leasing of its lands as attractive as possible.[1]

Some of the lessees who had been with the JAAP for years disgustedly dropped out when their grain leases ended. Part of the land they left was taken over by other old lessees who switched to grass and part by new people. Beef

---

[1]Intraoffice memo from Lloyd Shaid, Chief Engineer, Chicago District, U.S. Army Corps of Engineers, January, 1959.

cattle were (and are) brought in from as far away as Bloomington and Springfield in central Illinois to take advantage of what some consider the best sizable source of good, cheap pasture within several hundred miles. Per acre pasture rental consistently runs less than do taxes on similar privately owned land around the arsenal. A large portion of the hay land was leased by Prairie Creek Farms, an alfalfa pelletizing operation established specifically to take advantage of the abundant hay acreage the government was making available. Several nonrestricted field crops came up for consideration by land management officials as they sought ways for the government to derive still more income from the arsenal. Sweetcorn, broomcorn, and tomatoes were rejected mainly because they demanded the assistance of migratory workers who could pose security problems and who would definitely pose housing problems. Machine-harvested peas had done well at Badger Army Ammunition Plant near Madison, Wisconsin; but the JAAP climate is a little warm for peas during the podding season. Eventually, popcorn was chosen as the specialty crop at Joliet.[1]

Illinois popcorn comes primarily from counties in the southeastern part of the state; but a few Will County farmers had grown it, often as a curiosity crop, before the arsenal began advertising for bids on its five-year popcorn tracts in the early 1960's. Popcorn interest in Will County leaped from 170 acres in 1959, when it was still grown only on private land, to slightly more than 1,000 acres in 1969 (Table 15). Meanwhile, the balance of the state showed a decrease in acreage. Popcorn worked nicely into the grain farmer's cropping pattern, much as soybeans had done a couple of decades earlier. It requires no casual labor, and farmers can use most of their row-crop equipment and skills to produce it. Planting of popcorn can be postponed until after the field corn and soybeans are in. Harvesting can wait until late in the fall because, even though stalks droop, the ears remain firmly attached. And, unlike field corn, popcorn requires only modest amounts of fertilizer. As might be expected, however, there are some problems facing the popcorn producer--mainly in the marketing of his crop.

Popcorn producers, like vegetable producers, contract with processors who set the price per ton they are going to pay for the delivered grain. The popcorn farmer either agrees on the price before the seed goes into the ground or faces the uncertain, perhaps impossible, task of finding a buyer on the open market. In a year when quotas are not met by contract producers, the maverick

---

[1] Interview with Vernon Evans, Specialist, Real Estate Division, Chicago District, U.S. Army Corps of Engineers, April, 1969.

TABLE 15

POPCORN HARVESTED IN WILL COUNTY, ILLINOIS, AND IN THE OTHER 101 ILLINOIS COUNTIES: 1959 AND 1969

Year	Acres of Popcorn Harvested	
	Will County	Other 101 Counties
1959	170	18, 503
1969	1, 018	17, 652

Source: Illinois Agricultural Statistics, 1960, pp. 14-15; 1970, pp. 14-15.

grower might be able to sell his crop; but when there is a surplus of popcorn, he may be forced, as one arsenal grower was in 1968, to sell to the local elevator for a fraction of the market value. Failure to contract may even keep a farmer from planting any popcorn, because processors keep a tight hold on the preferred seed varieties which they then dole out to their contract holders in order to insure a steady flow of grain to their factory at harvesttime. Processors are located some distance from the JAAP so growers must bear a substantial transport cost to get their produce to market. Finally, as one who has popped corn might well imagine, artificial drying of this highly heat-sensitive material must be done with discretion.

Only a handful of Wilton farmers have actually leased Joliet Arsenal land. A number of others bid unsuccessfully during the boom period prior to 1956 when twenty or more men might have had aspirations on the same tract. Still another group of farmers admitted that they had contemplated bidding at one time or another but failed to do so. Some feared drawing the displeasure of their other landlords, some hesitated to take on the extra responsibility of a cash lease, and some could never see putting up with the inconveniences imposed by the installation on lessees. Just fourteen different Wiltonites appeared on the arsenal lessee lists in the six focus years since leasing began (Table 16).[1] No more than six of our farmers ever appeared at the JAAP in any single focus year. Many of the Wilton representatives rented plant land year after year, of

---

[1]In cases where the father began leasing and a son or sons took over later, we consider this a single operation.

TABLE 16

PERSISTENCE OF WILTON FARMERS AT THE JOLIET ARSENAL: 1944-1969

Lessee	Focus Years in which Wilton Farmers Leased Arsenal Land						
	1944	1949	1954	1959	1964	1969	Totals
A	x	x	x	x	x		5
B	x	x	x	x	x		5
C	x						1
D	x	x					2
E		x					1
F		x					1
G		x					1
H			x	x	x	x	4
I			x				1
J				x	x	x	3
K				x		x	2
L				x	x	x	3
M					x	x	2
N						x	1
Totals	4	6	4	6	6	6	

course. Lessee B, for instance, more than likely would still have been leasing there if he had not given up his land to avoid a conflict of interest when tapped as the new arsenal land manager in the mid-1960's. The crafty Irishman (Lessee A) and his sons who succeeded him were never really happy with the new popcorn, hay, and pasture program so they dropped off the list of active lessees between 1964 and 1969 after more than twenty years.

Arsenal rental has been an important source of nonfarmstead land for those from Wilton who have participated in the leasing program, and it has freed for the use of other farmers private land for which the fourteen might have successfully competed. Gradually, however, arsenal lessees are depending less and less on their government ground (Table 17). The sharpest drop in dependence occurred between 1959 and 1964 when the abrupt shift away from straight grain leases took place. About one representative popcorn grower's contemporary arsenal acreage, an astute observer of the leasing picture remarked:

> His popcorn is supplemental now to the beef feeding operation they've started. He could do without the popcorn, but the profits from growing it sent him and his wife to the dog show in Madison Square Garden this winter so it must be worth something.

Obviously, others must think arsenal land is worth something, too.

Thus far in Chapter V we have been discussing methods employed by

TABLE 17

IMPORTANCE OF JOLIET ARSENAL LAND
TO WILTON LESSEES: 1949-1969

Focus Year (1)	Total Arsenal Land Rented by Wiltonites (2)	Total Acreage of the Wilton Lessees (3)	Col. 2 as a Percent of Col. 3 (4)
1949	1, 940	3, 564	54
1954	1, 539	3, 231	48
1959	1, 946	4, 253	46
1964	1, 504	4, 108	37
1969	2, 444	6, 591	37

farmers in Deer Creek, Jackson, and Wilton to secure rights to nonfarmstead land. In the balance of the chapter we will direct our attention toward a particular and significant segment of nonfarmstead land--that which lies away from the farmstead tract.

The Matter of Tract Location

Having thought about how he might acquire more nonfarmstead land, the midwestern farmer must consider the characteristics of his present holding, of other land that is available, and of those tracts that could conceivably become available to him in the near future. To most potential expanders, the location of a tract with respect to the farmstead, the nearest town, or a good road is a characteristic of no small moment. A contiguous tract, that is a tract lying adjacent to the farmstead tract or connected to it by others that are adjacent (and controlled by the farmer in question), is often automatically more attractive than one that is not contiguous.[1] Despite their shortcomings, which we will discuss at length later, noncontiguous acres outnumbered contiguous nonfarmstead acres in 1939 and overwhelmed them in 1969 (Table 18). Both Jackson and Wilton in 1969 had five times as many noncontiguous acres as they had contiguous nonfarmstead acres while in Deer Creek the large number of farms helped pull the ratio down to 3. 5:1. As a percent of all township land, the non-

[1] No two tracts are contiguous unless they share a common property boundary or corner. Tracts that have between them a stream, road, or railroad are contiguous if the farmer and his animals can cross these barriers.

TABLE 18

LOCATION OF NONFARMSTEAD LAND WITH RESPECT TO THE FARMSTEAD, BY TOWNSHIP: 1939 AND 1969

Township	Year	Number of Nonfarmstead Acres that Were	
		Contiguous with the Farmstead Cluster	Not Contiguous with the Farmstead Cluster
Deer Creek	1939	3,651	4,791
	1969	5,225	18,766
Jackson	1939	2,028	4,909
	1969	3,006	16,563
Wilton	1939	2,164	4,620
	1969	4,075	20,146

contiguous portion has shown a fairly steady increase from around 20 percent in 1939 to 50 percent or slightly less in 1969. The proportion of farms partially composed of noncontiguous land has naturally increased, too, along with the noncontiguous acreage.

A Look at the Linkage Patterns

Many of the authors mentioned earlier who dealt with noncontiguity employed maps in their reports. Diller, for instance, simply assigned a number to each farmer and placed that number on all the appropriate parcels in his map.[1] Smith used an elaborate system that required numbering the farmer and his tracts, spotting his farmstead on the map, noting the acreage in any outliers and distances from the farmstead on the map, and connecting the farmstead and outliers with arrows.[2] To be absolutely safe, Smith also listed each farm in an appendix, repeating some of the data presented on the maps but noting a few other attributes as well. Aiken, since he depicted only one operation, could afford to draw a fairly detailed map showing each parcel being used by the Presley Plantation in 1970.[3] All systems have their good and bad points. Numbers on a map permit eventual identification of operators but fail to convey the feeling

---

[1] Diller, Farm Ownership, Tenancy, and Land Use, p. 178.

[2] Smith, "Road Functions."

[3] Aiken, "The Fragmented Neoplantation," p. 45.

of connectivity. Arrows between farmstead and outliers stress the connectivity but make operator identification practically impossible except at a fairly large scale. On the other hand, with large-scale maps, the cartographer encounters the problem of what to do with a few distant parcels. A map of the individual farm unit permits analysis without static from all the other tracts that are actually interspersed with those in this unit. There is a limit, however, to the number of farms one can depict in this manner.

We have chosen here to emphasize the gross pattern of spatial linkages at the expense of parcel size and shape or actual road distances and routes involved (Figures 9, 10, and 11). Further, we have depicted the linkages only in two of the focus years, although data were also gathered for the intermediate five, in order to save space and because the tremendous increase in complexity between 1939 and 1969 strikes us more if we have only the two years to compare. To highlight the linkages, the maps have been shorn of all but the essentials. Only the direct lines between farmsteads and points of access to outliers appear. Some idea of scale is obtainable from the six-mile-square township outlines, but the reader should remember that farmers can seldom travel directly from home to another plot unless it lies in one of the cardinal directions from the farmstead.

Maps drawn for the same year of different townships have more in common than maps drawn of the same township in different years. Farmers in 1939 were seldom on the road and of the small percentage who were, many had only a nearby outlier. From the 1939 maps, it is possible to distinguish without much difficulty every connection between farmstead and noncontiguous tract. By 1969, however, the tenure landscape has become such a blur that it defies even the most persistent to distinguish the separate connections. Smote by the apparent random confusion of the 1969 maps, we can easily conjure up visions of harried farmers rushing about trying to remember where they are supposed to be farming or where they left their yearling steers.[1] Perhaps, on the other hand, all is not as confused as it seems at first blush. It is possible that farmers might have considered their land's location as carefully as they did its other attributes but decided the advantages outweighed the disadvantage of not having it next door. The initial impression of chaos that one gets from the 1969 linkage maps will be tempered in the following pages as we search for the order that seems to prevail just beneath the surface.

---

[1]Occasionally, this author encountered farmers who during the interview temporarily overlooked tracts they were farming.

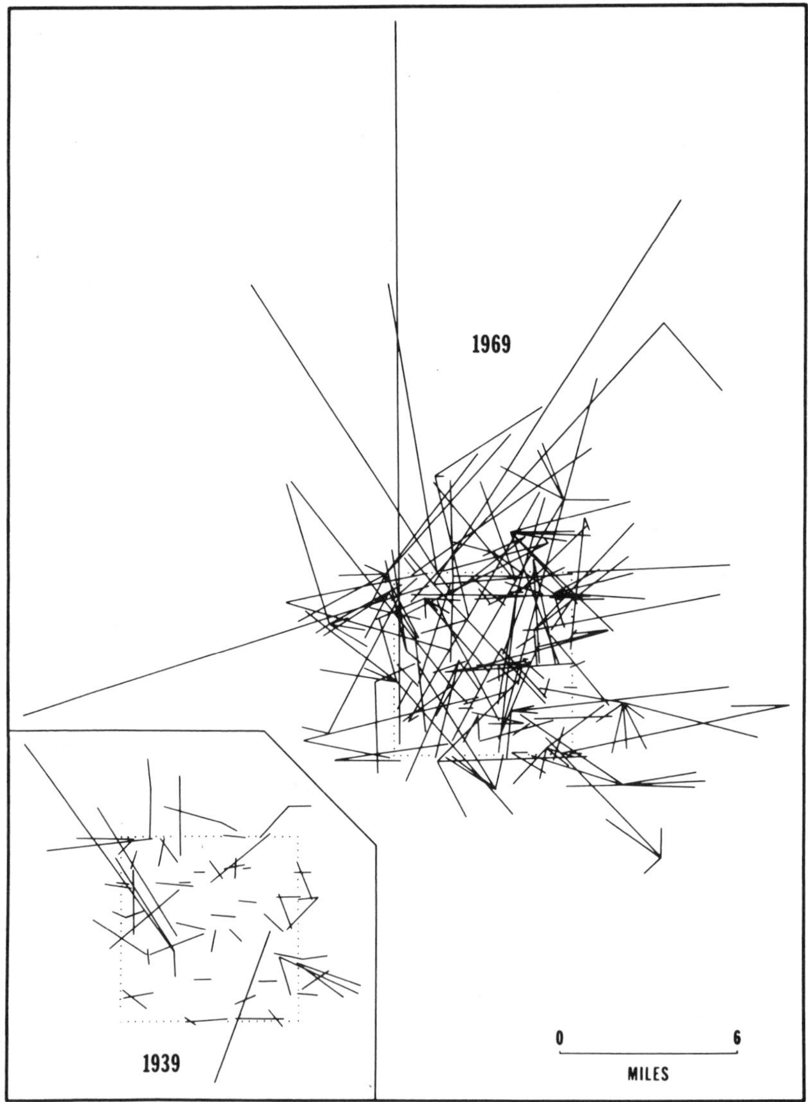

Fig. 9.--Tenure Linkages of Deer Creek Farmers: 1939 and 1969. Each linkage line is drawn cross-country between the farm headquarters and an outlying parcel in the operating unit. The township outline is also shown.

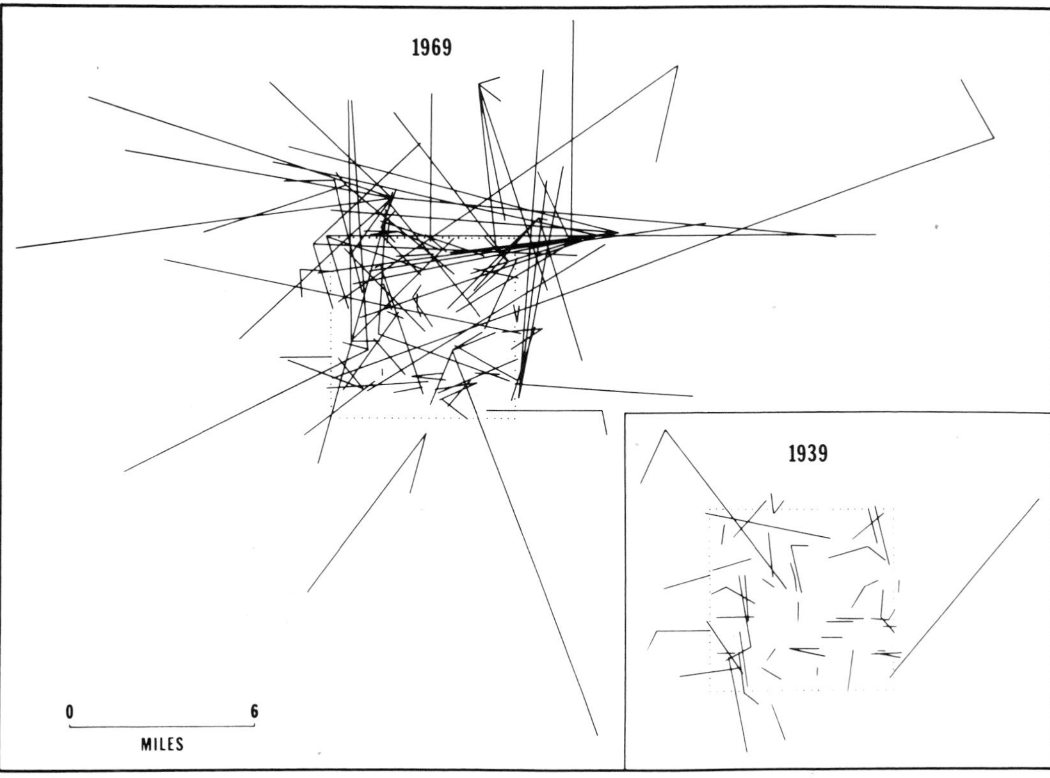

Fig. 10.--Tenure Linkages of Jackson Farmers: 1939 and 1969. Each linkage line is drawn cross-country between the farm headquarters and an outlying parcel in the operating unit. The township outline is also shown.

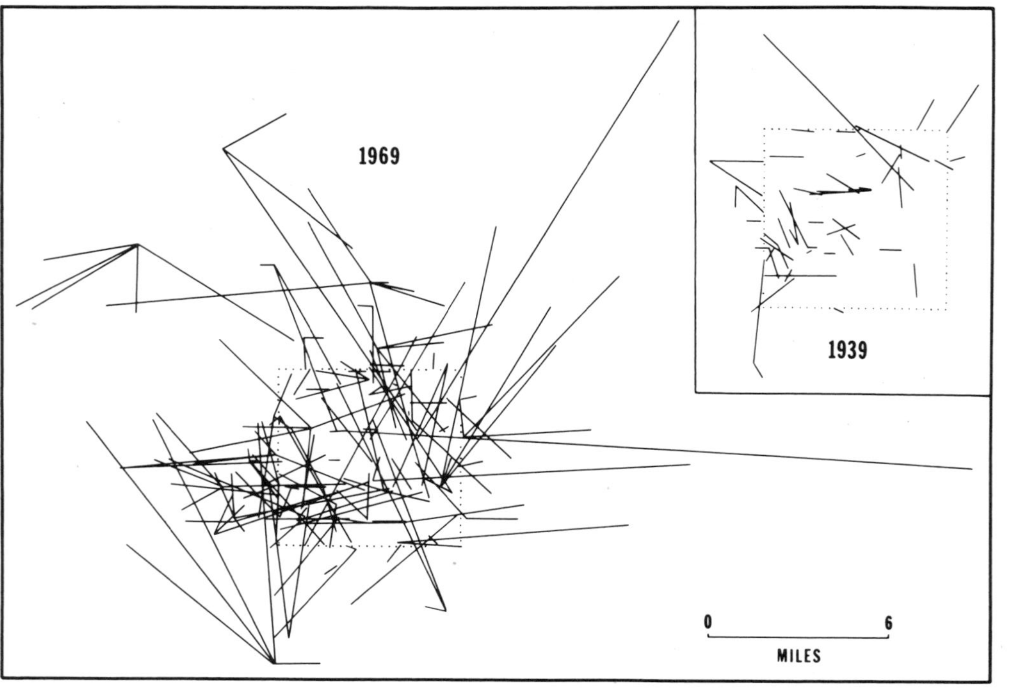

Fig. 11.--Tenure Linkages of Wilton Farmers: 1939 and 1969. Each linkage line is drawn cross-country between the farm headquarters and an outlying parcel in the operating unit. The township outline is also shown.

Distance to Noncontiguous Tracts

Although a farmer with a compact unit sometimes ventures out on the road with his machinery in order to reach a remote field, the man with a noncontiguous tract is faced with the necessity of transporting paraphernalia and personnel to it and from it several times annually. A noncontiguous tract of land may be more attractive to a farmer in many respects than those tracts which surround his farmstead; but against these attractions, he must weigh the time he will lose in transit and the money it will cost him to move. As Van Arsdall and Elder, observers of the contemporary Illinois agricultural scene, put it:

> locational fragmentation necessarily results in operational inefficiencies resulting from higher transport costs from headquarters to work areas and increasing difficulty of worker supervision.[1]

Large farms, they go on to say, are particularly susceptible to these cost and supervision difficulties; because, almost inevitably, expansion sooner or later means noncontiguity.

Distances used in this chapter were determined by measuring, with a wheel-planimeter, the shortest available road route between farmsteads and outlying tracts. Although farmers sometimes fail to choose the shortest route, its employment here facilitates comparison of distances in recent years, when routes can be recalled, with those of earlier years, when memories of actual routes taken have become a little fuzzy. Every reasonable effort has been made to eliminate from consideration routes that were unavailable to roving farmers because a bridge had washed away or a road had been abandoned. Men farming private land on opposite sides of the Joliet Arsenal must go around rather than through it. Even those farmers who lease arsenal land may find it necessary to use a circuitous route within the plant to reach their arsenal tracts. Admittedly, it would have been much simpler to have devised a method of estimating distance as did Sitwell in Nova Scotia and Nordbeck in Sweden,[2] but we have demurred mainly because we were curious about the actual distances. Moreover, during the three decades, Jackson distances were modestly altered by the construction

---

[1]Roy N. Van Arsdall and William A. Elder, Economies of Size of Illinois Cash-Grain and Hog Farms, Illinois Agricultural Experiment Station Bulletin 733 (February, 1969), p. 48.

[2]Oswald F. Sitwell, "Land Use and Settlement Patterns in Pictou County, Nova Scotia" (unpublished Ph.D. dissertation, University of Toronto, 1968), pp. 73-74; Stig Nordbeck, "Computing Distances in Road Nets," Papers of the Regional Science Association 12 (1963): 208.

of new, state-financed farm-to-market roads that penetrated the township from Highway 136 on the north and Highway 5 on the east.

While average farm size in the three townships doubled between 1939 and 1969, the average distance traveled by farmers to all their outlying tracts increased three to four times, if we consider just those who had noncontiguous tracts, and a whopping eight to ten times, if we include all township farmers (Table 19). There are now twice as many noncontiguous parcels in Jackson and three times as many in Deer Creek and Wilton as there were at the commencement of the study period[1] (see Column 3, Table 22). Thus even if the median distance to outliers had continued to hover around one mile as it was doing in 1939, total township travel would have ballooned anyway. But instead of hovering, each township's median distance to its tracts lying away from the farmstead nearly tripled. Farmers in Jackson, as an example, had twenty-four tracts in 1969 that were farther from their base of operation than the most distant Jackson tract had been in 1939. In fact, we need only add the mileages to the township's six most distant 1969 tracts to arrive at a sum that exceeds the 101.8 miles counted by the whole township in 1939.

TABLE 19

MEAN DISTANCE TRAVELED TO REACH ALL OUTLYING LAND FOR THOSE FARMERS WHO TRAVEL AND FOR ALL TOWNSHIP FARMERS, BY TOWNSHIP: 1939 AND 1969

Township	Mean Distance to Outlying Land from Farmstead in Miles			
	For Travelers		For All Farmers	
	1939	1969	1939	1969
Deer Creek	2.0	8.4	.6	6.2
Jackson	2.2	11.3	.8	7.7
Wilton	1.7	8.4	.6	6.5

What, we might wonder, is the likelihood of ever being able to exceed a township's 1939 mileage with just the distance to the single most distant tract in a given year? Figure 12 leads us to believe the chances are slim because in none of the three townships is there a persistent tendency for the maximum distance to increase. In Jackson and Wilton, the maximum to an outlier was reached

[1] Adjacent outliers are considered one tract.

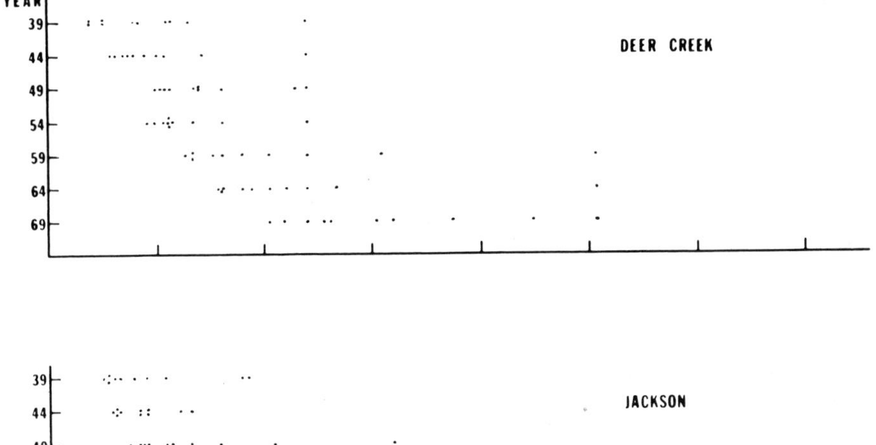

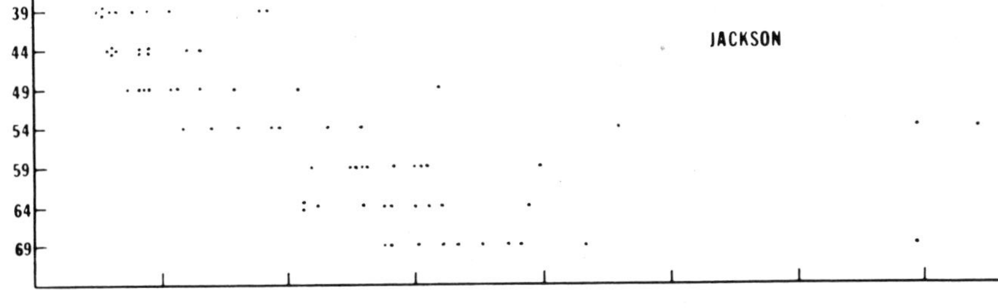

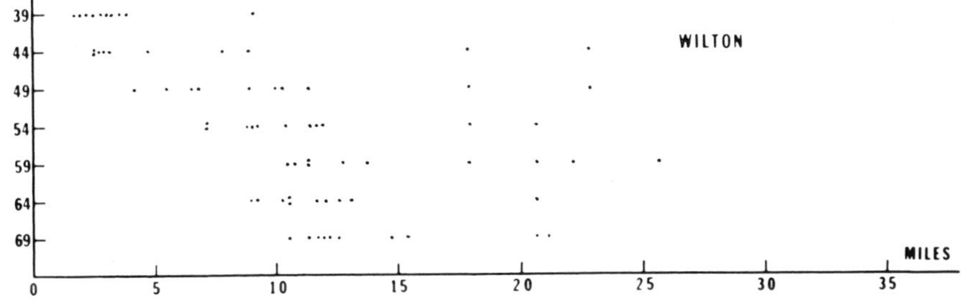

Fig. 12.--Road Distances to the Ten Most Remote Parcels, by Township: 1939-1969.

not in 1969 but in some earlier focus year. In Deer Creek there is currently an upward trend, but the last jump occurred a decade ago. Few farmers in the three areas have ever gone farther than twenty miles to farm. For those who have, with the exception of a pair of Jackson brothers (1954 and 1969), twenty-five miles seems to have served as a fairly firm barrier.[1]

Tract Size and Distance

Other things being equal, it stands to reason that a farm operator would be willing to travel farther for a large tract than he would for a smaller one. If he has to go to all the trouble of moving equipment or animals a considerable distance, then the acreage at the end of the journey should justify the extra effort expended getting there. Actually, farmers seem to have paid scant heed to size of outlying tracts when deciding how far they would go (Table 20 and Figure 13). While it is true that we consistently find a positive relationship between size of outliers and distances to them, in most instances the correlation coefficients are far too low even to bother mentioning the possibility of a significant statistical relationship. From the fact that Deer Creek and Jackson had considerably higher r values in 1939 than in 1969 we might surmise that tract size was once somewhat more meaningful to farmers when it came to obtaining noncontiguous land. Alas, we fail to find confirmation in the Wilton situation.

TABLE 20

SIMPLE CORRELATION BETWEEN ACREAGE OF NONCONTIGUOUS PARCELS AND ROAD DISTANCE FROM THE FARMSTEAD TO THESE PARCELS, BY TOWNSHIP: 1939 AND 1969

Township	1939	1969
Deer Creek	.33	.16
Jackson	.46	.17
Wilton	.07	.11

[1] Could there be a connection between this apparent barrier and the fact that the contemporary tractor can travel only a little more than twenty miles per hour in road gear?

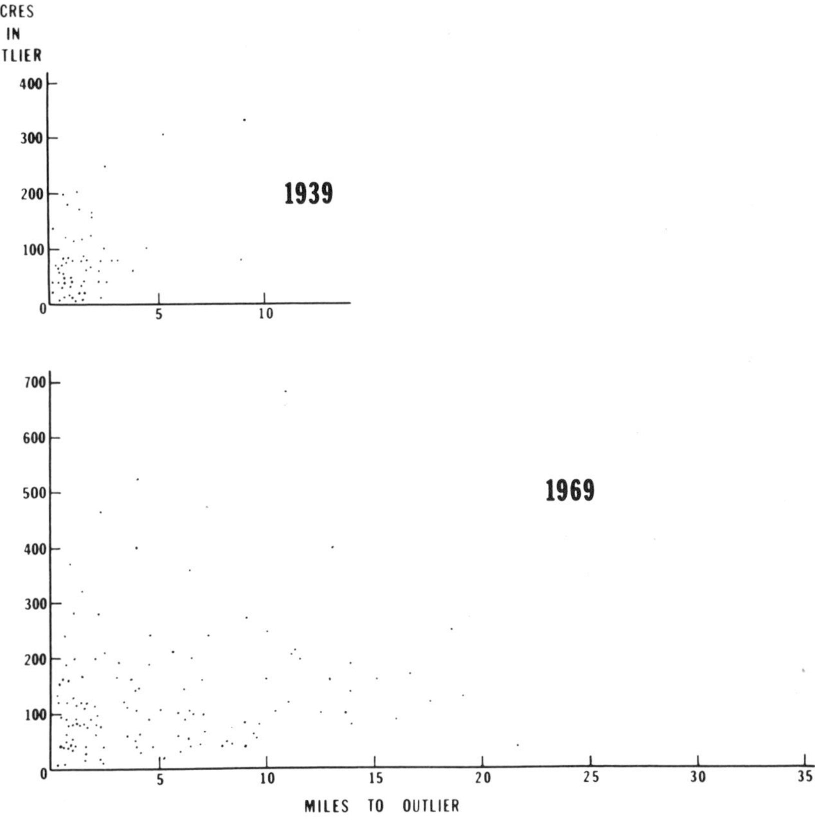

Fig. 13.--Relationship between Acreage of and Road Distance to Outlying Jackson Tracts: 1939 and 1969.

Size of Selected Tracts and Distance

We found little correspondence between outlier size and distance to the outlier, but what would happen if we were to look only at tracts in certain size categories? To test this notion, a 1939 list and a 1969 list of distances were drawn up for 40-acre tracts, 80-acre tracts, 160-acre tracts, and 240-acre tracts in each of the three townships. Because the government's surveyors left tracts that were slightly larger or smaller than these ideal sizes and because cemeteries, churches, nonfarm dwellings, and other land uses have a way of reducing a tract from a perfect eighty to 78.5 acres, for instance, a 5 percent deviation from the ideal was permitted. Therefore, tracts between thirty-eight and forty-two acres appear in the same category with true forties. In a couple of cases, the lists contained more than forty entries, but usually the number was less than twenty. Since no group was used if the number of eligible tracts dropped below three, we were forced to ignore 240-acre parcels in 1939 for all three townships. Once the distances had been arranged on the remaining twenty-one lists, it was a simple matter to determine the median distance for each category and prepare Table 21.

TABLE 21

MEDIAN DISTANCE TO TRACTS OF A SPECIFIED SIZE, BY TOWNSHIP: 1939 AND 1969

Township	Acreage of Tracts	Median Distance (in miles) from the Farmstead to Tracts of a Specified Size	
		1939	1969
Deer Creek	40	1.0	1.9
	80	1.1	2.6
	160	1.2	4.2
	240	a	2.4
Jackson	40	1.0	2.0
	80	1.7	1.5
	160	1.9	3.0
	240	a	7.3
Wilton	40	0.9	1.7
	80	1.0	2.5
	160	1.7	3.7
	240	a	2.6

[a]Fewer than three tracts in group.

Considering the dismal outcome of our previous experiment with size and distance, the results here are rather encouraging. In 1939, farmers were willing in all three areas to go a little farther for eighty acres than for forty and still farther for a quarter section. Despite some differences, Deer Creek and Wilton in many respects have more in common with one another than either has with Jackson. And nowhere is this similarity between our Illinois and Indiana representatives more noticeable than in the 1969 portion of Table 21. Just as it had in 1939, median distance traveled by Deer Creek and Wilton farmers increased as tract size increased from 40 to 80 to 160 acres. Beyond 160, however, median distances to the 240-acre tracts fell off sharply, back to the eighty-acre level. It almost seems as if somewhere between 160 and 240 acres we have reached a point of diminishing returns. In the eyes of Deer Creek and Wilton operators, 160 acres may have been the ideal size for an outlier--large enough to justify substantial road travel but small enough for convenient tending.

Jackson farmers, on the contrary, apparently had no inhibitions about acquiring distant 240's. They went more than twice as far for them in 1969 as they did for 160's and three times as far as did farmers in Deer Creek and Wilton for their 240's. Two possible explanations come to mind. First, since nearly all of the land in the five Jackson 240's was used for pasture, the need for machinery transport occurred less often than in the other two areas where the bulk of the land was cropped. Second, land in the five Jackson tracts came into the hands of its operators by purchase or cash rent. Either way, you eliminate the concern of the share landlord who might prefer to have a resident tenant or nearby nonresident farming his land. Diminishing returns do set in eventually. If we lump together all Jackson tracts with acreages greater than those in the 240-acre range, median distance drops back to 4.0 miles. Thus, Jackson is not so different from the other two townships after all. Instead, lower land quality and a different agricultural emphasis push a bit higher the acreage at which distance becomes a burden.

Land Quality and Distance

An unsuccessful attempt was made to discover the effects of land quality on the distance farmers will travel to farm. Actually, the analysis of quality never really got started because of our inability to perfect a reliable but simple-to-use method of determining the agricultural worth of a given unit of land. Several possibilities were explored and rejected, including soil productivity ratings, assessed valuations, and yield ratings assigned by Agricultural Stabilization and Conservation Service.

Land Use and Distance

Thwarted in our attempt to relate land quality and the distance traveled by farmers to outlying tracts, perhaps we can circumvent the problem by employing land use data as an indirect measure of quality. In Deer Creek and Wilton virtually all the privately held land, no matter what the quality, is devoted to the production of row crops; thus little is to be gained by examining land use patterns in these two areas. The Jackson situation, fortunately, is quite different. In Jackson there is a sharp distinction between land normally in grass and land suitable for serious cultivation. It is true that here and there one finds corn on the hills and hay in the bottoms; but, on the whole, use patterns seem to reflect land quality. For comparison we have divided Jackson's outlying land into two categories--pasture and hay in the first, annual crops in the second. Along with the pasture and hay we include acres diverted from corn because (1) these are mowed like hay and (2) farmers are commonly permitted to pasture diverted acres after a stipulated date in the fall. Since timberland in Putnam County provides livestock with winter shelter, summer shade, and some sparse forage, these acres also go into the grass category. Annual crops consist mainly of corn and soybeans, but there are occasional fields of wheat, sorghum, or oats.

In quest of land, Jackson farmers in 1969 traveled almost twice as far for every 1,000 acres of cropland (72.7 miles) as they did for the same amount of grassland (42.7 miles). Or put in another way, they were willing to move the same distance for one acre of cropland as they were for two of pasture or hay. Think for a moment what this means. We are not talking here about dairy cows that remain in pastures near the milking parlor so that the farmer can handle them personally twice a day but instead about beef cattle that fend for themselves between occasional visits by the farmer to count them and see that they have no special problems. Hay land, it is true, must be attended to with mower, rake, and baler once or twice a season; but the popular round (cylindrical) bales often stay right where they were dropped by the baler or end up stacked under plastic out in the field. Either way the cattle can be brought to the hay, thus saving the cost of transporting it back to the barn.[1] So, despite the rather unintensive nature of hay and pasture use, grassland is actually tied closer to the farmstead than cropland. What about the cropland? Aside from the few fields temporarily in corn each year prior to being reseeded, there are only limited areas where crops can be grown in Jackson Township. Often these are scattered, but if needed badly enough, farmers will make the extra effort to reach them. We conclude that in Jackson, at least, there is a causal relationship of sorts among distance, land use, and land quality. On an acre-for-acre basis, the better land (cropland) encourages more travel than the poorer (grassland) variety.

---

[1]Livestock can be brought in only if the farmer owns the hay tract or makes special arrangements to rent the pasture as well as the hay.

Special Circumstances and Distance

Size and agricultural value per acre are just two of a tract's characteristics that might influence a farmer's decision to farm or not to farm it. In this section we are going to introduce a series of special circumstances, any one of which could induce a man to reach out farther than he would if his only consideration was simple land hunger. Our intention is to demonstrate that while hunger for more land is often important, there frequently is something else about a tract that lures a farmer far from home.

1. Rental tract owned by a relative.

   No matter where it is, this type of land is difficult for a farmer to refuse. He seldom has to compete for it, he usually knows its attributes intimately, and if necessary he can briefly ignore it at rush seasons without fear of losing it to another renter. On the other hand, the related farmer might feel obligated to the owner (perhaps a widowed mother) to farm it even though other available land is more suitable or more suitably situated.

2. Rental tract owned by a primary landlord other than a relative.

   The tenant farmer treading precariously between the furrow and the assembly line knows there are certain landlords that must be satisfied with his performance if he is to remain in farming. From a primary landlord he might rent his farmstead, a critical portion of his land, or valuable land near the farmstead. If a primary lord also owns or manages other land, the operator in question could get a chance to farm it. To keep a primary lord happy, the tenant might take on what will be for him an inconvenience. Conversely, the success the tenant has enjoyed on the original land bodes well for him in negotiations for other tracts even though they are far away.

3. Rental tract located in a familiar distant area.

   Beyond a few miles from his farmstead, a farmer's knowledge of the land and potential landlords' knowledge of him decline rapidly in many instances. When the parties involved in the rental of a distant tract are neither related nor already involved in a leasing arrangement, they may have gotten together because:

   a. the farmer retained that tract when he moved to the present base. Just as urban families may depend on old friends for a while after moving to a new neighborhood, farmers at times deem it prudent to keep some familiar land, even though now inconvenient, while they establish their reputations in the new area. The most distant Wilton tract in 1939, for instance, was one held over like this. After a year or so the outlying land was given up, and the new Wilton farmer never rented away from home again.

   b. the farmer exploited his goodwill in a distant area to acquire rental land there. Certainly not the only goodwill source but definitely one of the most important is the family. If parents live near a tract, the owner is much more likely to look favorably on a request by the son to farm it than he would if he and the family were strangers. When the time comes to perform the farming operations, it also is beneficial for the son if his parents live close to the distant outlier. They can check on field conditions and livestock for him, and their farmstead is always available as a haven when he gets caught by bad weather while working nearby.

4. Rental tract proximate to another tract in an operation.

   Let us suppose a farmer already has a tract ten miles from his base. If another tract only a mile from that first tract comes up for rent, this farmer could be in a good position to compete for it. First, the sort of work he does is on display at the tract he already has. Second, the farmer must make the trip out into the area anyway; so unless the combined acreage would constitute a burden, it might be advantageous to augment what is already his. For our purposes, a tract is considered proximate to another if they are separated by a road distance that is less than the median distance to tracts for that year and township. Only one of the two tracts can qualify, of course.

5. Rental tract passed down from one relative to another.

   Leases in the Midwest do not pass between family members as commonly as they do in Britain, for instance; but such transfers do occur, thus giving a special advantage to the recipient.

6. Rental tract at the Joliet Arsenal (Wilton, 1969, only).

   The special land at the JAAP attracts farmers from considerable distances for reasons that were explained earlier.

7. Inherited tract.

   A farmer is not obligated to farm land he inherits but usually does. It is often land that he has farmed before or is farming at the time of the owner's demise. Even more important than his familiarity with the land is the simple fact that he probably needs the additional acreage. Like the rental tract owned by a relative, it is difficult to refuse.

8. Purchased tract--extenuating circumstances prevail.

   This category accommodates purchases for which reasons other than land hunger played a major role in the decision of the farmer to buy. Perhaps a man buys a tract out on the main highway or near town intending to move his farmstead there in a couple of years. Just because the tract is located at some distance from the present base of operations does not mean the farmer is going to move machinery and livestock to and from it indefinitely. This category will also catch tracts that were bought because they were near another outlier or came along as part of a multiparcel purchase.

The median distance to outlying tracts, for a given year and township, was used to distinguish distant tracts from those located in the immediate zone where travel is considered less of an inconvenience. Admittedly, this was an arbitrary way to separate the two groups; but no better method was available. The results are summarized in Table 22. Land hunger was in all cases a more potent force for the acquisition of land in the immediate zone around the farmstead than it was beyond the median. To put it another way, farmers were more likely to have a special reason for going out several miles than they were for

TABLE 22

PRIMARY REASON FOR INCLUSION OF NONCONTIGUOUS TRACTS IN OPERATING UNITS, BY TOWNSHIP: 1939 AND 1969

			Primary Reason for Inclusion in Operating Units			
			Land Hunger		Other than Land Hunger	
Township (1)	Year (2)	Total Tracts[a] (3)	Number of Tracts Located within Median Distance (4)	Number of Tracts Located beyond Median Distance (5)	Number of Tracts Located within Median Distance (6)	Number of Tracts Located beyond Median Distance (7)
Deer Creek	1939	61	17	13	13	18
	1969	209	56	21	48	84
Jackson	1939	58	12	5	17	24
	1969	124	35	32	27	30
Wilton	1939	50	10	6	15	19
	1969	160	38	16	42	64

[a]The few tracts right at the median distance have been excluded here.

traveling shorter distances. This tendency is least obvious in Jackson (1969) where the land-hunger category received a big boost from several distant pasture tracts rented in the late 1960's for no apparent reason other than the need for grass.

Operator Age and Distance

A farmer's health, age, wife, financial standing, community reputation, political connections, ambition, and a dozen other personal characteristics could conceivably affect the distance he travels for noncontiguous land; but most of them, except age, were impossible for this author to measure. Could there be a nexus between an operator's age and the amount of traveling he does in a given year? Might not young farmers wander more than their elders? Ages of 1969 farmers were determined during the interviews. In the few cases where a man refused to give an interview or to give his age during the interview, the present author estimated it after consulting written sources such as the man's service record or after soliciting assistance from friends and relatives who might know. For the ages of 1939 farmers, it was possible in many instances to work backward from the age the man gave at the interview. If a farmer had died, the gravestone, a newspaper obituary, or a relative could often provide the necessary figure. Age estimates were made only when available information seemed reliable. Otherwise, we deleted the farmer from this particular analysis.

Each farmer falls into one of five age cohorts: 29 and under, 30-39, 40-49, 50-59, and 60 and older. Using these cohorts, two tables were prepared. The first, Table 23, reports the percent of 1969 farmers in each age group who had outlying land. In all cases, the evidence points convincingly toward confirmation of our suspicion about youth and mobility. The youngest farmers are most likely to have non-contiguous land, and the oldest farmers, the least likely. Table 24 employs the same mileage-acreage index used earlier in this chapter in our discussion of Jackson land use. As a farmer's total acreage increases, no matter what his age, there is a tendency for mileage to increase also. By determining mileage per 1,000 acres and using that value, we eliminate the effect of total acreage on total mileage and improve our chances of learning whether age really has an effect of its own.

TABLE 23

PERCENT OF FARMERS WITH NONCONTIGUOUS TRACTS, BY AGE COHORT, BY TOWNSHIP: 1969

Township	Age Cohort	Percent of Farmers with Noncontiguous Tracts
Deer Creek	29 and under	92
	30-39	76
	40-49	75
	50-59	79
	60 and over	50
Jackson	29 and under	100
	30-39	87
	40-49	82
	50-59	68
	60 and over	35
Wilton	29 and under	100
	30-39	93
	40-49	85
	50-59	71
	60 and over	50

TABLE 24

MILES TRAVELED PER 1,000 TOTAL ACRES FARMED, BY AGE COHORT, BY TOWNSHIP: 1939 AND 1969

Township	Age Cohort	Miles Traveled per 1000 Total Acres Farmed	
		1939	1969
Deer Creek	29 and under	9.3	30.2
	30-39	2.1	18.4
	40-49	3.3	17.2
	50-59	2.8	22.2
	60 and over	0	11.4
Jackson	29 and under	3.2	16.3
	30-39	3.3	20.1
	40-49	5.3	15.0
	50-59	2.1	18.6
	60 and over	4.5	5.9
Wilton	29 and under	1.8	18.7
	30-39	2.5	16.5
	40-49	2.8	15.3
	50-59	2.3	18.3
	60 and over	1.1	5.0

The 1939 figures in Table 24 are generally unreliable and have been included only for the sake of comparison with those of 1969. So few 1939 farmers traveled away from home for land that one long-distance trek could completely dominate a cohort. This happened in Deer Creek to the 29 and under cohort and in Jackson to the 60 and older group. By 1969, a substantial proportion of the operators are farming away from home so that one farmer has difficulty dominating his cohort. The oldest group of farmers travels less per 1,000 acres than any other group. Many of these men are on compact parcels after dropping off outliers in their declining years, or maybe they never had them in the first place. Farmers under 30, as we anticipated, travel more than any other group in Deer Creek and Wilton. Travel for them is necessitated by the paucity of land near their headquarters, encouraged by their strong desire to "stay farming," and facilitated by the vigor of their youth. They need land, are willing to go after it, and are capable of enduring the extra hours on the tractor that reaching it demands. Jackson's youngest group in 1969 occupies only an intermediate position among the five cohorts. Each of the three Jackson farmers in the 29 and younger cohort did some traveling, but at the same time all have a surplus of land (mainly upland pasture) compared to the number of cattle they have to stock it. They are, therefore, more interested in building their herds than in getting more land.

The most unexpected Table 24 revelation must be the strong 1969 showing made by the 50-59 cohort. In Deer Creek and Wilton it is the second most mobile group, and in Jackson it is the most mobile. How do we explain such a consistent departure from what reason tells us should be the case? For one thing, the average farm size of men in their fifties is in all three townships smaller than the average size of farms operated by men in their forties. In two of the townships, the difference in farm size between the two middle-age cohorts exceeds 100 acres. Such could, of course, help account statistically for the strong position of the 50-59 cohort, but why are these farmers going so far for so few acres in the first place? This we do not know. Perhaps the reader would like to speculate.

CHAPTER VI

THE TROUBLE WITH NONCONTIGUOUS LAND

In the summer of 1959, an Ohio farmer named Jim Clark trucked his big diesel tractor and a specially equipped farm wagon to the New Jersey shore, unloaded them, hitched up, and pulled his family 3, 300 miles behind the tractor to the Oregon coast.[1] It was just something he had wanted to do since learning an ancestor of his had followed a somewhat similar route along with Meriwether Lewis early in the nineteenth century. To publicize the durability of a new tractor tire, Goodyear Tire and Rubber Company employees in the early 1960's drove a farm tractor pulling a loaded wagon 3, 689 miles from Maine to California. Farmers reading a subsequent Goodyear advertisement were asked to, "think how many trips to the field you'd make to equal that mileage."[2] Exactly ten years after Jim Clark tractored from New Jersey to Oregon, a seventy-six-tractor caravan of dissatisfied farmers drove 800 miles from Redmon in east-central Illinois through "the mud and mountains" to Washington, D. C., where they presented their ideas on farm prices to the House Agriculture Committee and to Illinois' Senator Everett Dirksen.[3]

If Jim Clark had followed ancestor William to Oregon in a station wagon, no one would have noticed. If Goodyear had driven just 3, 689 miles on a set of new passenger-car tires, they would not have bought a two-page spread to promote it. If the Redmon farmers had ridden chartered buses to Washington, fewer reporters would have bothered to monitor their excursion. The Farm Journal, however, felt readers would be interested in these three junkets be-

---

[1]Jim Clark, "Coast to Coast on a Tractor!" Farm Journal, September, 1959, pp. 64A and 64O.

[2]Goodyear advertisement, Farm Journal, March, 1963, pp. 24-25.

[3]Bob Coffman, "A 'Tractor March' on Washington?" Farm Journal, August, 1969, p. 26; Bob Coffman and Jerry Carlson, "What Those Tractor Marchers Did in Washington," Farm Journal, September, 1969, pp. 30 and 39-40.

cause they featured familiar farm machinery in an alien setting.[1] To take a tractor and implement out onto the public road is, as most farmers know, costly, time-consuming, and even unhealthy. In this chapter we shall discuss machinery movement and a number of other problems facing the midwesterner who, in the words of one elderly informant, is "farming up and down the road."

Extra Expenses

The thoughtful acquisition of noncontiguous agricultural tracts by American farmers has been described as "rational fragmentation."[2] Without much thought, farmers in Deer Creek, Jackson, and Wilton were able to provide this author with a long list of reasons why they would just as soon have all their land in one piece no matter how rational the fragmentation might seem to observers. For one thing, noncontiguity is expensive.

Unfortunately, nobody has bothered to find out just how expensive it can be. Jensen at Montana State indicated a desire to do so but, because of institutional funding problems, never got around to it.[3] Hays and his farm appraisal firm in western Indiana have developed a rule of thumb to guide them in estimating the damages due a landowner when a new highway is about to deny him access to a piece of land that was originally contiguous to his farmstead. They feel a five-mile separation raises the cost of cropping by 10 percent. Additional miles would cost him more than this but proportionally less per mile since terminal costs would remain fairly constant.[4]

Farm people, too, are aware of the extra expenses; but again nobody seems to know for sure how much they might be. A widow whose husband in the 1950's had traveled fourteen miles to a rented 120 acres in Wilton is convinced, "there's no money in traipsing around." A taciturn Wilton cattlefeeder admitted that even though he was temporarily farming two outliers, "it pays to be concen-

---

[1]Gathering data for this chapter and the next, the present author examined each monthly issue of the Farm Journal from 1939 through 1969.

[2]Gregor, Geography of Agriculture, p. 104.

[3]The study was proposed in: Clarence W. Jensen and Darrel A. Nash, Farm Unit Dispersal: A Managerial Technique to Reduce the Variability of Crop Yields, Montana Agricultural Experiment Station Technical Bulletin 575 (April, 1963), p. 18. Jensen's reason for not following up was explained in a: Letter from Clarence W. Jensen, Professor of Agricultural Economics, Montana State University, May, 1971.

[4]Hays, interview, Oxford, Indiana, March, 1970.

trated in one spot." Several Jackson farmers, when asked about noncontiguity, expressed their distaste for it in terms of the value they would place on an adjacent parcel. "To get it against me," said one ambitious young farmer, "I would pay $30-50 more per acre than the $150-200 it is worth." His older brother, who is farming essentially the same sort of land but with a more compact layout, would offer the seller a $25 premium. An older Jackson farmer whose sheep and cattle graze rather poor land along the eastern border of Jackson Township is ready to pay $15-20 more per acre if adjacent.

Time Expenses

Midwestern farmers, particularly those without livestock, have much idle time on their hands during the winter. Employers like the industrial contractor at the Joliet Arsenal have tapped this labor reservoir by hiring local farmers between December and March.[1] Lack of gainful employment is, however, not a problem for farmers during the growing and harvest seasons when the majority are actually hard put to complete field tasks on schedule. "Few industries," notes Hunt, "require such timely operations as does agriculture with its sensitivity to season and bad weather."[2] At no season is the farmer more vulnerable to the timeliness[3] factor than in the spring when, because of the weather and wet fields, he may have available just half the number of days shown on the calendar. On good planting days, a farmer's time is worth $50-100 per hour.[4] The more productive the land and the greater the capacity of the machinery he is using, the greater will be the cost to him of time wasted and of field operations performed after the optimum date.[5]

---

[1]Uniroyal, the firm holding the manufacturing contract at JAAP, was at first reluctant to hire farmers when reactivation of the plant occurred in the mid-1960's. They preferred to have year-round employees but quickly discovered local farmers to be much more reliable than workers lured into the area by the promise of high wages.

[2]Donnell Hunt, Farm Power and Machinery Management (Ames: Iowa State University Press, 1968), p. 3.

[3]"Timeliness is the measure of ability to perform a job at a time that gives optimum quality and quantity of product." R. B. Schwart, Farm Machinery Economic Decisions, University of Illinois Cooperative Extension Service Circular 1065 (December, 1972), p. 5.

[4]Kendle, interview, Kankakee, Illinois, January, 1970.

[5]In McLean County, Illinois, which is located at roughly the same latitude as Putnam and Cass but a hundred miles south of Will, corn yields on the aver-

How much of a farmer's time is spent in transit or transit-related activities? Take, for example, the case of a Wilton cash-grain farmer who was farming 557 acres in 1969. Besides his 160-acre farmstead tract, he owns and farms an eighty to the east which at its nearest point is 1.8 miles from his home. To the west he rents 200 acres, and due north of the 200 is another rental tract containing 117 acres. From the farmstead it is 1.6 miles to the 200 and 4.9 miles to the 117.[1] For this exercise we will assume it takes him ten minutes to get a machine ready for the road and another ten minutes at trip's end to put it back to work.[2] Tractors and self-propelled combines will travel 15, 18, 20 miles per hour and even a bit faster under ideal conditions but because of traffic and other delays a slower speed seems reasonable. We assume an average road speed of 12 miles per hour. This figure corresponds closely to actual time and distance estimates given this author by informants. The eighty was planted entirely to corn in 1969. The 200 to the west had 120 acres of corn and 80 of soybeans. To the north, the 117 was used for 30 acres of beans, 75 of corn, and 12 of winter wheat.

The 1969 crop year began after removal of crops from the preceding year (see Table 25). As soon as the soybeans were gone from the north 200 in the fall of 1968, our farmer disked down the stubble on twelve acres and drilled the wheat. Because the wheat went in away from home, the disking and drilling operations wasted 80 minutes each, including 30 minutes for preparations and 50 minutes en route. Timeliness in the fall can influence yields to a degree, but it is when plowing for the next year's crops that the farmer really begins to feel the pressures of limited time. To save valuable spring hours, many farmers

---

age begin to decline if the crop is not planted by May 5. For the period, May 5 to May 15, the decline is a modest one-half bushel per day. After May 15, the decline increases to one bushel per day of planting delay. For soybeans, the respective dates would be June 1 and June 10. Telephone conversation with Eugene Mosbacher, McLean County Extension Agent, Bloomington, Illinois, August, 1973. On the same topic Lantz wrote: "Experience has proved that corn planted about the third week of April in McLean County has the best yield prospect. And fields [farms?] have grown so large that to get most of the crop in on time, all field preparation has to be done and the planters running as soon as conditions permit in the first half of the month." Stanley Lantz, "Oak Leaves, Squirrels' Ears Determine Corn Planting Date," Pantagraph, April 15, 1972, p. B-10.

[1]When possible, he will catch fields at the 200 on his way to and from the 117.

[2]This is probably a conservative assumption.

TABLE 25

ESTIMATE OF TIME LOST BY ONE WILTON FARMER DUE TO NONCONTIGUITY: 1969 CROP YEAR[a]

Field Operation[b]	Crop[c]	Tract(s)	Lost Time in Minutes		
			Preparation[d]	Travel	Total
Disk Stubble	Wheat	North	30	50	80
Drag and Drill	Wheat	North	30	50	80
Disk Stubble	Corn & Beans	All	70	65	135
Plow	Corn & Beans	All	70	65	135
Field Cultivate	Corn & Beans	All	70	65	135
Plant	Corn	All	60	41	101
Plant	Beans	North & West	30	25	55
Rotary Hoe	Corn	All	60	41	101
Rotary Hoe	Beans	North & West	30	25	55
Cultivate	Corn	All	60	41	101
Cultivate	Beans	North & West	30	25	55
Combine	Wheat	North	30	50	80
Combine	Beans	West & North	50	50	100
Pick and Shell	Corn	All	70	65	135
Totals			690	658	1348 (22 hours & 28 minutes)

[a]We assume he follows a sensible circuit such as the one outlined in Chapter VII and that he is always able to finish an operation once started.

[b]Every farmer has his own tillage system. This one is fairly typical of Will and Cass counties.

[c]Crop acreage and tract location data based on interview with the farmer.

[d]Assume 10 minutes per preparation. An exception is the final storage at home for which no noncontiguity time is assessed.

[e]Assume an average speed of 12 miles per hour on the road.

now try to get as much plowing done in the fall as possible despite knowledge that this practice encourages wind erosion during the winter. To put off plowing until after the thaw, could mean trouble if it turns out to be (what some call) a backward spring. During the final seedbed preparations and planting period, time lost on the road can mean a lower yield in October. The farmer in our example is really rather lucky. While he spends the equivalent of three regular working days in transit, there are many in Wilton, Deer Creek, and Jackson who see a lot more of the public road from the seat of a tractor in a year than he did in 1969.

Machinery Expenses

Time is not the only expense incurred by the farmer who must move his machines from place to place. Fuel and lubricants have to be considered. Hard-surfaced roads accelerate tire wear. Rough roads can damage sensitive machine parts. And even though the farmer might be unaware of it, his insurance premium reflects the risk he and other men are taking when they venture out with their machinery into traffic.

Fuel and Oil

The fuel and oil used on the road in farm machinery constitute an unproductive expense. A young Deer Creek partowner had eight noncontiguous tracts in 1969 totaling 917 acres on which he planted nothing but corn. With a 21.9-mile circuit to the east he can catch five of the eight. To the south, a 9.6-mile jaunt enables him to work two more. Finally, he needs a 9.0-mile round trip west of his farmstead to take care of the last parcel. If his tractors and combine average 100 horsepower, he should use the equivalent of five gallons of diesel fuel per hour of general field work.[1] In road gear, under less strain, consumption will drop to about four gallons per hour. Minus state and federal motor fuel taxes, diesel fuel in 1969 cost him around 20¢ per gallon or 80¢ an hour while out of the field.[2] If our energetic farmer made seven visits to each of his eight tracts for a total of 284 miles and was able to travel at an average speed of 12 miles per hour, he spent nearly 24 hours between fields. The $19 worth of fuel and oil used moving would have plowed almost seventy additional acres.[3]

Tires

Farmers wasted no time adopting pneumatic tractor tires after their 1933 introduction. By 1939, the first year of our investigation and 100 years after Charles Goodyear first vulcanized rubber, more than half of the new agricultural tractors were rolling out of the factories on rubber instead of the tradi-

---

[1]Schwart, Farm Machinery Economic Decisions, p. 4.

[2]Technically, farmers are not supposed to claim a fuel-tax refund for those gallons burned during intertract treks.

[3]Assume a plow width of 120 inches and use Schwart's "Machine Capacity" equation to determine acres per hour for the plow. Schwart, Farm Machinery Economic Decisions, p. 6.

tional, nearly indestructible steel.[1] Since World War II, virtually all models have come equipped only with rubber. The advantages of rubber over steel were easy to see and easier to feel. Pneumatic rubber tires reduced rolling resistance and in many instances increased traction. It took less fuel to do a job. Machines and operators were jostled less because rubber reduced the transfer of ground irregularities through the wheels. Machines lasted longer and farmers suffered less after a day in the driver's seat. Vibration was exaggerated whenever steel-lugged wheels rolled over such unyielding surfaces as the public highways. This fact coupled with state laws forbidding the use of unprotected steel lugs on the hard roads helped keep most midwestern farmers at home with their implements. Pneumatic tires eventually altered this relationship between the farmer and the public road. Still the shift to rubber was not without its drawbacks.[2]

Rubber tires are expensive. To replace an original equipment rear tire with one of comparable quality, a farmer will spend about three hundred dollars; premium tires cost even more. The farmer needs at least two traction tires for each tractor plus two others for his combine. Furthermore, it is no longer the road that suffers in the encounter between it and tractor tires. Rubber lugs have always been quite susceptible to scuffing, or wiping, when driven at road speeds over concrete or blacktop surfaces. As one dispersed Irishman in Wilton put it, "Blacktop cuts the tires off our tractors. More lug rubber is lost over a mile of blacktop than in a week of field work." From a Deer Creek-area operator came a similar complaint: "Even at three-quarter speed, with ballast and a heavy implement behind, you can almost see the tires wear out on pavement. If a farmer does much road work, he will need to replace tires every two years." A Jackson farmer, who also runs a tire agency, maintains that if a tire is properly used and kept off the road, it will rot out or burst its sidewalls due to age before a farmer can wear off the lugs. High road speeds also affect tires in a more subtle way. According to a B. F. Goodrich informational brochure, high speed "creates excessive heat under the tread which weakens rubber and cord fabric. Although damage may not be visible at the time, premature failure may follow."[3]

---

[1] "Magic Carpet," Saturday Evening Post, December 31, 1938, p. 22.

[2] F. Hal Higgins, "Rural Revolution on Rubber," Pacific Rural Press, August 26, 1939, pp. 110-11.

[3] What You Should Know about Farm Tires (Akron: B. F. Goodrich Tire Company, n.d.), p. 15.

Among our three townships, Deer Creek's farmers ought collectively to be losing the most lug rubber on the road. For one thing, the sum of their one-way distances to outlying tracts in 1969, 761.1 miles, is greater than Jackson's 680.2 or Wilton's 620.7. For another, more roads in the Deer Creek area are now hard surfaced than in either of the other two townships (see Table 26). Of the 88.0 road miles in and along the margins of Deer Creek, 77.5 are surfaced with some type of blacktop.

TABLE 26

ROAD MILEAGE, BY SURFACE TYPE, BY TOWNSHIP: 1969[a]

Township	Surface Type						Total Township Miles
	Hard[b]		Gravel		Dirt		
	Miles	Percent[c]	Miles	Percent	Miles	Percent	
Deer Creek	77.5	88	10.5	12	0.0	0	88.0
Jackson	9.3	19	25.3	51	15.0[d]	30	49.6
Wilton	18.3	24	53.0	68	6.3	8	77.6

[a]Includes roads shared with adjacent political units.

[b]Primarily blacktop. U.S. 52 through Wilton is the only concrete stretch of any significance in the three townships.

[c]Percent of total township road mileage.

[d]Includes some gravel stretches that have been allowed to deteriorate to the point where mud makes them impassible.

The Machine Itself

Even if alternative routes are available so that a farmer can avoid tire-shredding blacktop and concrete, he may prefer to use them and re-tire every other year rather than expose his expensive and often sensitive machinery to washboarded gravel or heavily rutted dirt roads. "Road gear is hard on machinery," commented a former Jackson farmer now working in Kansas City. "Tractors are poorly sprung and the chuckholes can shake things loose." Manufacturers of farm equipment try, if possible, to reduce the likelihood of rough-road damage by using stronger materials than they would if the machinery always stayed in the field.[1] Naturally, these additional costs are passed on to the farmer in the purchase price. Also passed on are costs incurred by the manufacturer in making it easier and safer for the farmer to take his machines away from the field and onto the road. Portability devices like hydraulic folding mechanisms plus such safety items as flashers, flags, and slow-moving-vehicle emblems could be eliminated if farmers were able to accumulate enough land in one block.

Impediments to Machinery Movement

As farms grow larger, farm implements grow, too. A big farmer needs and demands big machines to do his work as quickly as possible during those crucial periods when field conditions are ideal or at least tolerable. What happens when he puts these big machines on roads designed with the automobile or even the team and wagon in mind?

Gates

In the first place, the farmer may have to make special arrangements just to get through his farmstead gate. We often hear about perimeter fences being torn out to make room for another row of corn next to the ditch, but farmers have been modifying access points in their fences for years as new and wider implements were acquired. Narrow gates can be especially onerous when farmers find them impeding ingress to otherwise attractive rental parcels. Is the parcel worth the expense of modifying someone else's gate; or could it be farmed profitably with those old, discarded implements rusting away out behind the barn? They might be less efficient in the field than their replacements, but at least they would go through the gate.

Roads and Shoulders

Once past the gate, a farmer and his machinery will be traveling mainly over roads that may or may not be wide enough for two cars to squeeze past one another. Along Jackson byways, especially the 30 percent that are still dirt, overhanging brush will reduce clearance even more and force the operator to travel cautiously so as to avoid damage to protruding parts of the equipment and his person. Shoulders (berms to many Hoosiers) are usually too narrow to

---

[1] Interview with Robert Thayer, National Service Manager for Portable Elevator Company, Bloomington, Illinois, July, 1973.

accommodate a machine if the farmer pulls over to give another vehicle the right-of-way. Alternatives include taking to the ditch, a practice which invites minor damage to the machine or even an overturn, or staying on the roadway and forcing the other driver to find his own way around. In the words of the largest grain farmer with land in Wilton during 1969:

> My equipment and I spend a lot of time on the [heavily traveled] Peotone-Wilmington Blacktop. It is foolish to keep pulling off onto those narrow shoulders every time a car wants to pass. You lose time and you might lose a bundle if you happened to pull off at the wrong place.

Solid waste has a way of collecting along rural roads. A discarded ironing board can be as upsetting as a weed-concealed gully or gravel pile. A Wilton farmer complains about road signs erected by county highway department employees in the middle of the shoulders to warn motorists of upcoming dangers (see Plate III). On several occasions, he has been forced to halt at these signs with his implements and wait until the way was clear before pulling out and around them.

Farmers moving north and south on Indiana Highway 29, a mile west of Deer Creek Township, encounter a rather historic hazard just south of Deer Creek (the stream). As noted earlier, Indiana 29 is a descendent of the Michigan Road. In order to keep the Road open across the floodplain of Deer Creek, it was decided in the nineteenth century to surface it with logs. Sycamores were

Pl. III.--Rear view of a "Do Not Pass" sign located six feet from a new blacktop road in Manhattan Township, Will County, Illinois. This and similar signs on the road shoulder can impede the movement of wide farm machinery.

chosen, cut to the appropriate length, and laid side by side to form a corduroy road. According to the legend, sycamore saplings sprouted at the ends of the logs and grew to form what the natives call Sycamore Lane or The Sycamores.[1] At places, the 300-yard-long corridor is barely wide enough for anything else when farmers brave enough to use it start through with an implement wider than ten feet. Fatal accidents have prompted highway authorities to institute tree-removal campaigns from time to time, but preservation groups like the Carroll County Historical Society are attempting to resist them.[2]

Country Bridges

To the typical motorist, load and size limitations of the bridges he uses are of little moment because his vehicle is normally well below them. To the midwestern farmers, however, bridge capacities are a source of constant concern. Not only are supplies trucked to and products hauled away from his farm, but bridges must also permit passage of implements destined for outlying parcels of land. Farmers give far more thought to the weight of trucks and school buses than they do to that of their field equipment. Usually spanning less than forty feet, most bridges in and around the three townships easily accommodate a 10,000-pound tractor and its 4,000-pound implement. Likewise, overhead clearance is seldom a problem because most country bridges have no overhead bracing.[3] As the reader may have guessed by now, the crucial bridge parameter for dispersed midwestern farmers is width. It is to this problem we will address ourselves in the next few paragraphs.

Bridges in the three study areas are simple in design with most employing the same rigid-beam principle. Many of the oldest country bridges in the rural Midwest are made primarily of metal (see Plate IV). A set of girders under the floor and perhaps a pair of metal trusses on either side of the roadway bear the weight of the bridge and its users. To provide a floor, rough planks are laid side by side parallel to the stream banks. Sometimes two wooden

---

[1]"Legend Surrounds Deer Creek Trees," Pharos-Tribune, April 12, 1950, p. 1.

[2]At one time, the Society was planning to request that the trees be placed on the National Register of Historic Places. "Sycamore Lane Plans Told at Carroll County Session," Pharos-Tribune, May 27, 1970, p. 8.

[3]Underpasses can be troublesome, however. A Wilton farmer reported having to drive his combine through downtown Joliet on U.S. 30 to reach a tract he had rented west of that city. Low underpass clearances made alternate routes impossible.

runways are laid the length of the bridge over the planks and nailed down. Virtually all metal bridges in these areas date from the years leading up to and including World War I. Built of steel or wrought iron, they are long-lived if protected from erosion and traffic abuse.[1] After World War I, bridge authorities turned more toward steel-reinforced concrete as a substitute for the metal and wood bridge. These could be installed faster and perhaps cheaper. Also since the floor is made of concrete, there are no boards to rot, wear, or wash away with high water. Many of the concrete structures in Wilton and Deer Creek were erected at the time the roads they serve were upgraded from dirt to gravel (see Plate V). The sturdy sides of a concrete bridge may serve some support function but are primarily there to keep vehicles from sliding off the edges.

In the opinion of many farmers, their interests would be better served if something was done to eliminate bridge sides altogether or at least to push them farther apart. As a matter of fact, progress is being made by public agencies responsible for bridges; but there may never come a day when roving farmers will gain complete freedom from bridge harassment. Some farmers have resigned themselves to taking long, time-consuming detours. Others completely avoid land from which they are separated by such a bridge. Still others choose their new implements carefully with one eye on field efficiency and another on the bridge situation. "I had to get an Allis-Chalmers combine instead of the Massey-Ferguson I really preferred," recalled one Jackson operator with bridge problems, "because the smallest model that Allis offered was a [crucial] foot narrower than the smallest in the Massey line." An Irishman in the southwestern corner of Wilton had no special bridge complaints but did admit passing up a wider combine last time so as to avoid trouble: "We [he and his son] didn't feel a two-foot wider head would cut enough extra grain to make up for the time we would lose removing it in order to pass our problem bridges." For a farmer determined to cross a bridge with equipment that in the field is too wide, there are ways. In the next chapter these methods and devices will be among our primary concerns.

Much grumbling about bridge widths is heard from farmers in Jackson and vicinity, a bit less from southern Will County, and almost none in the Deer Creek area. Table 27 suggests a simple explanation. Of the thirty-seven bridges measured in Deer Creek and surrounding townships by this author, 89

---

[1] For an affectionate look at the vanishing iron bridge east of the Appalachians see: William Edmund Barrett, "Iron Bridges," Pioneer America 3 (January, 1971): 21-32.

Pl. IV.--A narrow pony truss bridge along the Wilton Township-Peotone Township line south of the Peotone-Wilmington Blacktop. The trusses on a pony truss bridge are so low that any overhead bracing would prevent a person from riding erect across it on the back of a pony.

Pl. V.--Poured concrete bridge located 1.2 miles west of Wilton Center, Illinois. Although built in 1937, at twenty-four feet the Jones Bridge is one of the widest in the three townships.

TABLE 27

PERCENT OF COUNTRY BRIDGES OPEN TO IMPLEMENTS OF A CERTAIN SIZE, BY TOWNSHIP: 1969

Width of Implement in Feet	Percent of Bridges that Were Open		
	Deer Creek	Jackson	Wilton
20-23	11	4	7
19[a]	22	4	15
18	24	4	17
17	46	4	22
16	51	4	22
15	89	8	51
14	95	20	81
13	97	60	98
12	100	68	100
11	100[b]	100[b]	100[b]

[a]Might include implements measuring as much as 19'11". The same goes for subsequent widths.

[b]We measured 37 bridges in the Deer Creek area, 25 in Jackson, and 41 in Wilton.

percent are open to implements in the fifteen-foot width range. For Jackson, only two of twenty-five are as obliging to traveling farmers. Furthermore, while virtually all the Deer Creek and Wilton structures are wide enough to permit passage of four-row equipment (about thirteen feet), fully 40 percent of those serving Jacksonians are too narrow. Not all Jackson farmers, however, suffer equally from their turn-of-the-century bridge network. Most severely affected are those ambitious men who want either to crop big bottomland tracts along the creeks or to custom combine (mainly soybeans) for other farmers. As one older cattleman put it, "Bridges don't give me any headaches, but you ought to hear my two young neighbors hollering about those down on East Locust." We did.

"Traffic + Tractors = Trouble"[1]

In the United States only mining and construction are considered more dangerous occupations than farming.[2] Annually thousands of farmers, their

---

[1]Norval J. Wardle, "Traffic + Tractors = Trouble," Iowa Farm Science 15 (May, 1961): 10-12.

[2]Charles S. Floyd, "First Step in Engineering for Safety: More Information, Please," Implement & Tractor, March 21, 1970, p. 20.

relatives, and hired hands are killed or injured performing farm tasks. Many of the fatalities, perhaps nine hundred each year, involve tractors.[1] In any given year, about one-third of these tractor-related deaths will occur on the public roads despite the fact that implements spend less than 5 percent of their time on the road. "The chances of having an accident with a tractor on the road," writes Wardle, "are five times as great as in the field or yard. And if there is an accident, the chances of its being fatal are over eight times as great on the highway. . . ."[2] Here is a partial account of an accident that added one and maybe two persons to Missouri's list of 1969 road fatalities. It happened only a few miles west of Jackson Township:

> A 65-year-old farmer was killed here last night in a tractor and truck collision . . . five miles north of U.S. 136 in Mercer County. . . . A passenger on the tractor . . . was taken to the Children's Mercy Hospital in Kansas City with severe head and leg fractures. She was reported in critical condition today. . . . The accident occurred when a state highway department truck . . . came over a hill and struck the tractor as it was entering the highway.[3]

This northern Missouri accident may or may not have involved a noncontiguous parcel. Tractors are, of course, taken on the road for other reasons, too. There is no doubt, however, why the Iowa farmer described in the following selection was out in traffic:

> Farmer C was returning home late at night after disking a nearby field. . . . His tractor was equipped with one white light to the front but the only lights he had to the rear were reflectors on each end of the disk. A motorist came down the road behind him, saw the two reflectors and, thinking they were markers for bridge abutments, drove between them. As the car rammed into the disk, Farmer C fell in the path of the disk and died instantly.[4]

Accidents on the public road are not necessarily fatal to the tractor driver although he is often thrown violently from the seat or pinned under the wreckage.[5] Assume a farmer survives the accident (see Plate VI). What are

---

[1] U.S., Department of Transportation, Agricultural Tractor Safety on Public Roads and Farms (Washington: Government Printing Office, 1971), p. 15.

[2] Wardle, "Traffic + Tractors = Trouble," p. 10.

[3] "Crash Kills a Farmer," Kansas City Star, July 31, 1969, p. 4.

[4] Norval J. Wardle, Operating Farm Tractors and Machinery Safely and Efficiently, Iowa State University Cooperative Extension Service Pm-450 (March, 1969), p. 60.

[5] It is difficult to speak confidently about the number of nonfatal road accidents involving tractors in the United States. Fatal tractor accidents must be reported as such, but no states separate nonfatal accidents involving tractors

Pl. VI.--Aftermath of a nonfatal, head-on car-tractor crash at the crest of a central Illinois hill. In this instance, the car and its driver were damaged more than the tractor and farmer. (Pantagraph photo)

the other possible consequences? Injury to the tractor driver or damage to the machinery could force costly postponement of a field activity at a critical point in the crop season. In extreme cases it might even be necessary for the farmer to relinquish hard-won rental acres to his eager competitors. If the tractor operator is at fault in the accident, he (his father if he is a minor) or the insurance company could be held responsible for damage to the person and property of others.[1] Collisions between farm equipment and other vehicles, as well as near-misses, not only jeopardize the welfare of the farmers actually involved,

---

from those that do not. Department of Transportation, Agricultural Tractor Safety on Public Roads and Farms, p. 6.

[1] In a recent Nebraska Supreme Court decision, the negligence of a 12-year-old tractor operator who had been involved in a traffic accident was affirmed. His father and the insurance company were assessed almost $50, 000 for damages. Keller v. Wellensiek, 181 N.W. 2nd 854 (Neb. Sup. Ct. 1970).

they also can damage the public image of the agricultural community as a whole at a time when farmers can ill afford to be making enemies.[1]

Safety specialists attribute the hazardous nature of road travel by tractor to a variety of things--inadequate brakes, inadequate lights and warning devices, operator error. Not infrequently, however, the speed of the agricultural equipment is mentioned as a contributing factor both in collisions with other vehicles and in those road mishaps in which only the farmer is actually involved.[2] Traveling at 12-15 or even 20 miles per hour, a farm tractor might just as well be parked in the road when we consider the much greater speeds of those who will meet or overtake it (see Plate VII). To make matters worse, the tractor driver hears virtually nothing besides his motor, whining tires, and blaring radio. On the other hand, tractors in road gear may be traveling too fast for conditions. A tractor, especially if it is carrying rather than pulling an implement, will begin to bounce if speed is too great. "When this [bouncing] occurs," writes Schnieder, "it is easy for the tractor to go out of control."[3] Many farmers now partially fill the drive wheels with a special liquid or powder called ballast for extra weight and traction. At substantial speeds on the road, instead of remaining in the lower portion of the rolling tire, the ballast will start to carry over the top causing the tractor to bounce erratically.[4] On turns the probability of an upset is greatly increased unless speed is reduced to compensate for the effect of centrifugal force.[5]

Farmers in Deer Creek, Jackson, and Wilton had much less to say about the dangers of machinery movement than about other inconveniences and disadvantages. Naturally, precuations are taken; but farmers tend to pride themselves on their courage and ability to endure hardship. "If those nuts out on the Peotone Road don't get me," says one Wilton humorist, "then I could live indefi-

---

[1]For a lucid discussion of the declining political power of American agriculture see: Clawson, Policy Directions for U.S. Agriculture, pp. 373-80.

[2]Wardle, Operating Farm Tractors and Machinery Efficiently and Safely, p. 53.

[3]Rollin D. Schnieder, "Tractor Design and Its Relation to Safe Operation," in Department of Transportation, Agricultural Tractor Safety on Public Roads and Farms, p. A-24.

[4]Ibid.

[5]Wardle, Operating Farm Tractors and Machinery Efficiently and Safely, p. 45.

Pl. VII.--A slowly moving light tractor impedes the flow of faster traffic on a Sullivan County, Missouri, blacktop leading north to Jackson Township in Putnam County.

nitely." Since they often pay taxes (or rent) on the roads that border their land,[1] many feel they have as much right to the road as anyone else and are reluctant to relinquish the right-of-way even when their own well-being is at stake. The philosophy of this middle-age Jackson farmer seems fairly typical:

> There is only one gait when moving and that is high gear. On the highway [U.S. 136] where the bridges are just wide enough for two lanes, we go right down the middle. Between the bridges, you've got to hold your half of the road, stay off the shoulder, and never motion cars behind you to go around. In other words, look after your own interests.

---

[1]"No provision was made for road easements in the United States Land Survey, but . . . settlers were usually willing to release land for this purpose because of the advantages of proximity to a highway. Taxes are usually paid on such land, the center of the road serving as the limit where roads separate properties." Norman J. W. Thrower, Original Survey and Land Subdivision: A Comparative Study of the Form and Effect of Contrasting Cadastral Survey (Chicago: Association of American Geographers, 1966), p. 89. Recently some state and county authorities have begun to purchase rights-of-way instead of seeking easements.

Livestock Movement Problems

Our attention has focused thus far on machinery movement, but farmers may also wish to rotate livestock from tract to tract with the changing seasons or for other reasons. In Deer Creek and Wilton practically all animals that move go by truck or trailer even for short distances. The story that accompanied Plate VIII when it appeared in the Pantagraph explains:

> The cattle trailer pictured is also of interest. Designed especially for carrying show herds, Perring [the farmer] bought it instead for transporting cattle from one pasture to another. Driving cattle a mile or more can be difficult because so few fields have fences these days. It is hard to keep critters from wandering to corn fields on the way. . . .[1]

As farmers in Jackson acquire more remote pasture tracts, the amount of stock hauling increases; but trailing along the road is still quite common. It is not, however, without its inconveniences. Traffic and a few unfenced fields necessitate extra drovers. Steers are especially troublesome on drives because, unlike the herd cows who may have made a trip several times before, everything along the route is new to them. All types of cattle, it must be reported, have a tendency to balk when asked to cross Jackson's wooden-floored bridges.[2] "We trailed them home from a pasture over on Route 5," said a farmer living south of Jackson in Sullivan County, "and had a terrible time pushing them across the Locust Creek bottom. They almost scattered so next time we'll take the truck." Another farmer confessed that he and his helpers had "worked all afternoon when it was below zero getting the cows across to the place where we like for them to calve later in the spring."

Special Managerial Concerns for the Dispersed Farmer

Accessibility is only part of the problem facing a man with noncontiguous land. An outlier will almost certainly prove more difficult to manage than similar land near home.

Grain Tract

Condition of the Soil

Throughout the year, the typical farmer will know less about the condi-

---

[1] "LeRoy Bull Going to Brazil," Pantagraph, June 8, 1972, p. D-1.

[2] Farmers feel this reluctance to cross is the result of the cattle being able to see through cracks and holes to the water below.

Pl. VIII. --A Prize Angus bull poses in front of the trailer in which he will ride for the first portion of his journey from central Illinois to a breeding assignment in Brazil. The farmer also uses the trailer to transport less valuable stock from one pasture tract to another. (Pantagraph photo)

tion of his noncontiguous land than he would like. The spotty nature of growing-season rainshowers and thunderstorms means it could be too muddy to work at one place but just right five miles or even a mile away. "I must drive out there a dozen times a year in the pickup," said a time-conscious Jackson farmer, "checking the moisture content of the soil." Checking "is a waste of time and money," in the opinion of one Deer Creek operator, "but it beats pulling over with a tractor and implement only to find yourself unable to get in the field."

Equipment Repairs and Adjustments

No matter where a rear tractor tire fails, there is little a farmer can do but summon a serviceman to remove and replace it with a new one or a loaner while his is being repaired professionally. Many machinery problems, however, can be handled by the farmer provided he has access to the proper tools and, during inclement weather, a suitable place to do the job. Ordinarily, a farmer carries only a few small hand tools with him on the tractor or in the truck. Seldom would he have on hand a welding outfit, for instance, even though one would be available back in his warm, dry machine shed. A minor welding job that might take fifteen minutes could cost a half day of downtime just because it happens while he is away from home. Even more costly than the hours spent going back to the farmstead for repairs might be the permanent damage sustained if he should yield to the understandable temptation and continue using it.

Equipment Thievery and Vandalism

"You're plowing ten miles from home," hypothesizes the *Farm Journal*, "and it's getting late. So you drive back in the pickup and leave the tractor in the field hoping to get an early start next morning. Don't bank on it: Thieves Can Steal Your Tractor Tonight."[1] Thefts, of course, are not confined to outlying tracts; but far from the owner's farmstead, a tractor thief can work with less worry about detection than if he had to spirit it out of the barn. Thieves may also find grain stored in isolated cribs and bins more tempting than that which the farmer has harvested at home or brought back home from an outlier (see Plate IX). Soybeans are especially popular among thieves because they are worth much more per bushel than corn. A farmer might never miss a single truckload of grain; and even if he did, one bushel looks pretty much like another when it comes to identification.

Vandalism has been most bothersome in Wilton where farmers are beginning to feel the effects of the advancing suburbs. Pranksters in 1969 managed to activate a big tractor that had been left overnight on a distant tract along the Peotone-Wilmington Blacktop. After playing for awhile, they rammed it against the corncrib and left the motor running in low gear so that the rear wheels would slowly dig themselves into the ground. The repair bill was substantial not to mention the time lost by the farmer from field work. Anything not tightly held down or locked may be missing when the farmer comes back to work the next day. Fuel is siphoned out of tractor and combine tanks if kids can find no easier way to steal it.

---

[1]"Thieves Can Steal Your Tractor Tonight," *Farm Journal*, April, 1969, p. 23.

Pl. IX.--A wooden corncrib set in the midst of a central Illinois cornfield. Isolated storage sites such as this make the work of grain thieves a bit safer. (Pantagraph photo)

Livestock Tract

Regular Chores

Some types of livestock require massive doses of attention. Milk cows seldom venture far from the farmstead for this reason. Hogs and feeder cattle might be assigned to a distant tract, but most farmers prefer to avoid this practice unless they can acquire a better than average set of livestock facilities along with outlying land. The wife of a Deer Creek farmer inherited an eighty that had a good setup for hogs, so he moved most of his animals down to it. Annually, he will make 300 ten-mile round trips to the lot just for chores. Noncontiguous pastureland is, of course, used the year around in northern Missouri and southern Iowa for cow-calf herds. During the warmer months, the farmer's presence is not required except to provide salt and perhaps a chemical solution to help control face flies and other insect pests. Beef cows, a few of which may reach twenty-five years of age, learn to take good care of themselves and their offspring. Still a farmer feels he should check them once a week and perhaps count to make sure his are all there and no interlopers are eating his grass. In the winter and at calving time the need for attention increases sharply. Just keeping hay available at several outlying spots can be a burden particularly when dirt and poorly graveled roads are involved. Jackson's creeks and ponds are frozen solid during several winter spells. Holes must be chopped, usually on a daily basis, or efforts by the cows to do it themselves can lead to a collapse of the ice and a sick or drowned animal.

Strays

Seldom does an issue of the Republican appear without a classified advertisement similar to this: "STRAYED: A yearling steer from my pasture on Neal Martin Farm in Hartford. . . . "[1] Not only is this animal valuable in its own right, but the farmer knows he might be sued if it causes an accident, gets into neighboring crops, injures other animals, or spreads disease. Straying from the farmstead tract or other nearby land is less of a problem for Jackson cattlemen than is straying from an outlier. On the one hand, a man will just naturally check stock and fences more often near home. On the other, neighbors are more apt to report a stray to an owner they know than to a stranger from across the way who is renting the stalk field[2] next door for only a couple of months. "Most people won't call me to report my strays," says a big Jackson cattleman with land scattered all over western Putnam County, "unless their property is being affected." Another complained, "Those folks down by the creek seem to think it's funny when my cattle are out on somebody else."

Rustling

Livestock thefts have intermittently plagued farmers in the Midwest for many years,[3] but the task of the rustler has been simplified in recent years by the reduction in the number of people living in rural areas. Some rural communities have been so decimated by population loss or diluted by an influx of rural nonfarm folk that the informal sentry system which once served to protect property from outsiders has all but disintegrated. Uncommon are those still knowl-

---

[1] Classified advertisement, Republican, November 12, 1969, p. 6.

[2] After a field of corn is harvested, cattle may be turned in to salvage the grain that remains on the standing stalks or the ground.

[3] See for instance: W. F. Schuyler, "Rubber-Tired Rustling," Farm Journal, September, 1937, pp. 17 and 58.

edgeable enough about local farm affairs to stop a strange truck and ask the driver his reasons for loading up a herd. "No one is willing to pry anymore," mused a Jackson cattleman. The dispersed nature of many livestock operations cannot help but contribute to the theft problem. Some of the Jackson outliers have occupied buildings; many do not. Quite often a tract has become an outlier because of its remote location--nobody wants to live there. Cattle, on the other hand, do not seem to mind; and it certainly does not bother the rustlers. Consider the details of this story which appeared in the Republican late in 1967. A farmer in the northwestern part of Putnam had his cows and their calves grazing on an outlying tract. Rustlers herded some of them into a pen and loaded several, using the farmer's own loading chute. For some reason, twenty-one others were left stranded in the small pen. A number had died of starvation before the farmer's landlord happened to discover them while on a furniture-storage mission to the old farmhouse two or three weeks later.[1]

Theft of livestock in the sixties was not a serious concern of Wilton and Deer Creek farmers. The only cattle in Wilton to get far from the farmstead lots are those belonging to arsenal pasture lessees. Given the arsenal's strict security, chance of theft there is slight. Houses are more frequent in Deer Creek than in the two other townships and many are still occupied by farm-reared people. Thus, the Deer Creek security system seems pretty much intact.

---

[1]"Thieves Left 21 Head of Cattle to Starve," Republican, November 8, 1967, p. 1.

CHAPTER VII

EASING THE BURDENS OF NONCONTIGUITY

In this chapter we shall learn how farmers, their suppliers, and public officials help ease some of the burdens described in the previous chapter. The reader must bear in mind, however, that the remedies themselves are often expensive and thus not always available to every farmer.

What Farmers Have Done

Farm Layout

Substitution of a contiguous or nearby tract for a distant one does occur but less frequently than farmers would like. Few are as fortunate, for instance, as the Ohio farmer described in a Farm Quarterly article.[1] This man was able to "eliminate one great timewaster--rushing from farm to farm" by persuading his relatives, who owned most of the land he was renting, to sell out and buy land around his farmstead. Land swapping was suggested by von Thunen as a remedy for wasteful farmstead-field travel;[2] but when the present author mentioned it to midwestern farmers, their reactions were mixed. Some had given it serious thought, and one Deer Creek man had tried to convince the owner of a strategically located sixty to trade it to him for an excellent but more distant forty. The owner of the sixty, as it happened, was not interested.

Elsewhere in Deer Creek, a land exchange could be in the offing according to a neighbor who has been watching the manipulations. There are in this tenure scenario, two pairs of brothers. Pair A invaded the home territory of Pair B to buy an eighty. In retaliation, Pair B bought thirty-three acres of what was once Pair A's home place, squarely in the midst of the land now farmed by Pair A. Our observer, who hates to see a farm split up, would like Pair A

---

[1] Charles Chester, "From 7 Farms to 1," Farm Quarterly, Winter, 1963-64, pp. 84-87.

[2] Johann Heinrich von Thunen, Von Thunen's Isolated State, ed. by Peter Hall; trans. by Carla Wartenberg (Oxford: Pergamon, 1966), pp. 62-63.

to recover the thirty-three acres. Optimistically, he believes Pair B "came up here to buy that thirty-three so later they can use it as a lever to pry away the eighty down on Highway 29." Most farmers, we must confess, have never considered the possibility of voluntary land redistribution by means of trading. "We could never decide on the quality of our land compared to someone else's," they say, "and with rental land two additional persons, the landlords, would have to agree. No, it would never work."

Despite the reluctance of farmers to swap, consolidation of land into larger tracts did nevertheless occur in the past thirty years (see Table 28).[1] Median farm size, however, is increasing faster than that of either home tracts or noncontiguous tracts in each of the three townships. Part of this farm-size increase is attributable to consolidation, but a large portion is not. Farmers in 1969 just happen to have more separate tracts than they did in 1939.

When Machinery Leaves the Farmstead

Once he has assembled an operating unit for the year, there are a number of ways for a farmer to expedite machinery movement and improve his away-from-home efficiency. Several of these are discussed below.

Planning Ahead

Time of Move

Farmers can reduce the chance of accident and the number of right-of-way yields by carefully choosing the time of day and week for implement relocations. Night moves, for obvious reasons, are to be shunned. Several Wilton and Deer Creek farmers expressed concern over the use of their highways and farm-to-market blacktops when local factories like the Joliet Arsenal are changing shifts. "We do our moving during the middle of the day," said a Wilton farmer who has been going east 20.6 miles to his mother's 160-acre farm for over fifteen years. A Deer Creek grain farmer admitted the rush-hour traffic on U.S. 31 north of Kokomo bothers him some but quickly added, "I use the road when I get ready because I pay taxes on four acres of pavement." Early morning moves seem to be the preference in Jackson where there are no special

---

[1] As farm numbers decline, some consolidation is inevitable. The chances of contiguity are considerably greater, for example, when there are only 75 farmers in an area than when there were 150 in the same area. For a good discussion of this topic see: Smith, "Road Functions," pp. 178-84.

TABLE 28

MEDIAN SIZE OF FARMS, HOME TRACTS, AND NONCONTIGUOUS TRACTS, BY TOWNSHIP: 1939 AND 1969

Township	Median Size in Acres		
	1939	1969	Percent Increase
	Farms		
Deer Creek	126	252	100
Jackson	162	353	118
Wilton	178	380	113
	Home Tracts[a]		
Deer Creek	112	120	7
Jackson	140	220	57
Wilton	160	200	25
	Noncontiguous Tracts[b]		
Deer Creek	60	80	33
Jackson	70	100	43
Wilton	80	94	18

[a]Home tract includes the farmstead tract plus any nonfarmstead land contiguous to the farmstead tract.

[b]Noncontiguous tract can be a single outlying parcel or more than one if they are themselves adjacent to one another.

factory rush periods and where high-speed through traffic tends to increase later in the day. Several Wilton and Deer Creek men noted certain days are worse than others for traffic hazards. "I go where I please," answered one Deer Creek farmer, "except I avoid [Highway] 29 on holidays." Friday and Sunday afternoons are hectic along blacktops leading south from Wilton toward recreational areas along the Kankakee River.

Circuit

During the long winter months, farmers will make many decisions about the coming year. A man with more than one tract of cropland must give thought to his spring work schedule; and quite often in these deliberations, tract location is important. This is particularly true in Deer Creek and Wilton where the differences between tracts is not so great as in Jackson with its sharp upland-bottomland distinction. Ideally, the planting sequence for a farmer in our Indiana and Illinois townships would go something like this. He begins at home, planting corn.[1] When finished, the planter is moved to the nearest outlier and from there to another and another in a logical, predetermined pattern. Road travel and preparation time are minimized by finishing each job before moving on and by taking each tract as it comes. When the last of the corn is planted, the farmer may be several miles from home--perhaps, but not necessarily, at the most distant tract. A few minor adjustments are made in situ, and the farmer is ready to backtrack through the circuit toward home planting soybeans. Rotary hoeing and cultivation can now follow the same sequence of tracts--out with the corn and in with the beans.

At harvesttime, because of their shorter growing season, the beans will be ready first. Most farmers prefer to begin cutting them at home to make sure the combine is functioning properly. Then they either take the familiar tract-by-tract route or go to the far end and come back. After the beans are out, the combine returns to the farmstead for maintenance and changeover from the grain platform to the corn head. Because farmers like to finish picking corn at home,[2] they may go directly to the most distant cornfield first.

There are, as might be expected, innumerable reasons why the ideal Wilton and Deer Creek circuits are not always followed. Weather or poor drainage can force the farmer to abandon the circuit just to keep working. Few farmers enjoy leaving a field only partially done, but they will if a lengthy delay seems likely.[3] Landlords can upset the best-laid schemes. "I try to keep him

---

[1]One Wilton farmer's lease stipulates he must plant all the 232 acres of his farmstead tract to corn before venturing away to outlying land.

[2]By the end of corn harvest in late fall, the weather has usually deteriorated to the point where farmers will need a comfortable building nearby even for routine maintenance and adjustments.

[3]A dispersed farm unit can be an asset during periods of patchy precipitation. Agricultural economists working with wheat yield records in the Great Plains reported some farmers will obtain widely spread tracts in order to minimize the damage potential of a single hailstorm. Don Bostwick, Studies in Yield Variability, Montana Agricultural Experiment Station Bulletin 574 (January, 1963), pp. 38-47; and Jensen and Nash, Farm Unit Dispersion: A Managerial Technique to Reduce the Variability of Crop Yields. In the Midwest it is not the hailstorm but the shower and thunderstorm that are the dominant types of summer precipitation. A farmer with all his land in a single contiguous block could be delayed several days while his neighbor who has other land scattered around the township could be out working as he waits for the farmstead tract to dry up.

[the owner of a distant eighty] happy," said a small Deer Creek operator, "by planting some corn on my farmstead, moving over to plant all his, and then coming back here to finish." The plight of one big Wilton cash-grain farmer struck this author as amusing. This individual farms 1,119 acres in Florence[1] and Wilton townships, including four rented tracts belonging to his mother, aunt, or sister plus an eighty belonging to him and his brother. The planting sequence he prefers is similar to the one outlined above. "I like to drop my planter in here at home first before taking to the road," he said, "but then my mother started griping because I didn't get to the farm where she lives until a week or so later. Oddly enough, she owns both places." His sister gives him trouble, too. "She lives in Arizona, but is always telling me how I should be getting better crops here in Illinois. I've already quit farming her land twice." To keep everybody happy, he went on, "I use the circuit but start one farm ahead each year and always do my eighty last."

Duplicate Sets of Certain Implements

Few farmers can afford to own all the equipment they would like, but some are helping themselves by obtaining extra implements specifically for outlying tracts. Seldom are these duplicates seen at the farmstead. The Deer Creek farmer who for a number of years has traveled twenty-five miles to a place north of Logansport keeps an old combine, a disk harrow, and a planter there so that he will not need to move them. The Wilton farmer mentioned earlier who goes east to his mother's farm maintains a core of tillage equipment there and roads only the planter, grain drill, and combine. Another Wiltonite says he would not have driven the twenty-five miles back into Kankakee County in 1959 for 166 acres if the landlord had not furnished a small tractor and selected implements. Duplication, however, is not confined just to men facing extremely long treks. A Deer Creek farmer has bought and deposited old wheelless disks 3.8, 2.9, and 2.1 miles from his home. "Everytime I get another farm," he says, "I go to an auction and buy a disk. Add a place, add a disk." Some farmers also like to have duplicate fuel tanks at their different farms. If the outlier has a set of inhabited buildings, the occupants can help watch the tank for him. If not, a secluded spot such as that shown in Plate X should be selected.

---

[1]Florence is the first township west of Wilton.

Pl. X.--Elevated fuel storage tank located away from the road on an uninhabited outlying tract in Putnam County.

Getting Assistance

Farmers are a self-reliant lot; but when it comes to moving machines and coordinating cropping operations at distant parcels, most will need help. To cut down reconnaissance trips, a favored few can depend on the firsthand advice of people living on or near their noncontiguous tracts. A telephone call to a farm-reared informant is often all a farmer needs in order to decide whether or not to road over his machinery and go to work.[1] Even if the person knows nothing about agriculture, he or she can at least report if last night's thunderstorm affected the field in question.

Farmers obtain assistance in other ways. The young Deer Creek farmer whose fuel usage on the road was discussed in Chapter VI employs a man just to keep the tractor crews rolling during the busy planting season. Hauling every-

---

[1]While the author was visiting the widow of a man who had rented Deer Creek land in 1939, her tenant (a son-in-law) called to ask if the beans were ready for the rotary hoe. Since her husband had never owned such a tool, she refused to speculate and told him to come and see for himself. When seated again at the kitchen table, she noted his dependence on the phone and on her as his informant.

thing from fertilizer to drinking water, the backup man roams from tract to tract seeking to minimize the avoidable delays. Most farmers, however, depend on their families to perform these supportive chores. Wives, children, retired fathers and fathers-in-law may be asked to work nearly as hard as the farmer in planting and again at harvest.

Allis-Chalmers, in a 1939 advertisement for its pull-type combine, tried to convince the wives of potential buyers that they, too, would benefit from the purchase of this new implement. The combine was going to make threshing and the hungry threshing gangs obsolete so there would be "no [more] slaving over a hot stove for Mother."[1] Now farm women are expected not only to cook the meal but also to deliver it to wherever the men are working. This the farm wife does in addition to running back and forth to town for supplies to keep the seed hoppers full and the chemicals flowing. "I'm in the pickup half the day sometimes," said one young Jackson wife who seems to enjoy those frenetic weeks.

Although too young to drive the car or pickup, farm children can be turned loose on the road with a tractor and implement in Indiana, Illinois, or Missouri at any age provided they are working for their family. "The kids [a boy and girl] move the tractors while I go to town for seed," said one Jacksonian. Naturally, most farmers exercise discretion when entrusting a child with a job having such serious possibilities. "When my ten-year-old boy goes on the road with us," another Jackson farmer noted, "we make sure he's pulling something narrow enough to stay in his lane of traffic. The hired hand or I always stick with him, too."

Retired farmers whose sons or sons-in-law are farming nearby find their services in great demand during April and May as backup men and again in the fall to help haul crops out of the fields. Experience and interest (often financial) in the crop outcome make them valuable team members. An additional advantage accrues if they happen to live on one of the outliers and can help coordinate activities at that end. The Irish father of a four-brother partnership farming south of the Joliet Arsenal must have had a busy schedule trying to meet all their needs before he died in the mid-1960's. One of the sons has now turned to his father-in-law who recently retired from a Wilton farm ostensibly to relax.

---

[1] Allis-Chalmers advertisement, Farm Journal, February, 1939, p. 29.

Neutralizing Narrow Bridges

With a Cutting Torch

When steel railings on minor bridges serve no structural function, bridge-plagued midwestern farmers are sorely tempted to eliminate them. In a couple of minutes, a man and his cutting torch can sever the vertical angle irons at the point of contact with the bridge floor and neatly drop the railing in the draw. "We torch a lot of bridges around here," confessed a man who once farmed in Jackson but now lives near Milan. The practice is, of course, illegal without official sanction. Before reworking the south railing on a branch bridge near the center of Wilton Township, the sponsoring farmer received permission from the road commissioner (see Plates XI and XII). They wisely made provision for swinging the railing up to its normal position, when needed for motorist protection, by bolting the uprights back onto the stubs.

With a Hammer and Saw

The stoneboat or slide has been used for centuries to haul rocks out of fields. Now, to save several extra miles of road travel each year, two Jackson farmers have found another use for it--dragging long, narrow field implements across narrow bridges. Built of heavy lumber, the slide shown in Plate XIII is perhaps four feet wide and sixteen feet long. Once brought to the bridge area, it remains alongside the road indefinitely, waiting for those few annual occasions when the builder (or another traveling farmer) needs it. The implement is pulled onto the slide and secured. The tractor is unhitched from the implement, rehitched to the cable at the end of the slide, and driven across the eleven-foot pony truss bridge. On the far side, the process is reversed. Altogether, the interruption may consume ten minutes.

With a Bulldozer

At scattered spots exasperated midwestern farmers have resorted to dozing out fords beside perfectly good but too-narrow bridges. Almost any ford, no matter how suitable the location, will cost money and take time to maintain. Often a ford is impractical because the approaches cannot be sloped back enough for safe entry and exit or because the stream would silt them up too quickly. Several Wilton-area farmers benefit from and plan to help maintain the ford pictured in Plate XIV. It was developed over a period of years by the farmer who operates land at both ends of the thirteen-foot-wide bridge located just across

Pl. XI. --A January view of a narrow Wilton bridge. The south (left) railing (as explained in text) pivots to this position to permit wide implement use.

Pl. XII. --A March detail of the bridge shown in Plate XI.

Pl. XIII.--Homemade implement slide lies beside a Jackson Township road and near a pony truss bridge over a tributary of East Locust Creek. See text for explanation.

Pl. XIV.--A narrow Kankakee County bridge and the ford (left) used by farmers to bypass it. Note the similarity between this pony truss bridge and the one in Plate XIII. Wilton Township is on the horizon.

the line from Wilton in Kankakee County. Originally, no noncontiguous land was involved; but during the late 1960's, other farmers, moving to their outlying land along this or nearby parallel roads, received permission from the owner to use it in lieu of the bridge. Because of the heavier traffic, the owner has added a gravel surface to make the ford more permanent. The other users will gladly supply future loads of gravel, not as a toll but rather as a gratuity for a man who is doing them a great favor.

Hauling and Towing Aids

In the 1950's, always alert for useful around-the-farm gadgets, the *Farm Journal* began publishing photographs and brief descriptions of transport devices built (or commissioned) by farmers to ease their noncontiguity burdens. Most observers, including perhaps the *Farm Journal*'s editors, apparently had felt after the introduction and rapid acceptance of pneumatic tractor tires that it was only a matter of time until the need to load implements on trailers for transport would disappear. Mounted on their own rubber tires, implements would follow wherever the tractor went. Yes, but (1) farmers were much slower to switch to rubberized implements than to rubberized tractors and (2) even when they did get around to converting or trading off old tools for rubber-mounted models, there were still plenty of tools that for years after World War II came without any wheels at all. Furthermore, the expansion of farms to the point where the farmer would need to take a utility vehicle to the field with him so that he could make quick trips to town or back home for lunch caught everyone except the farmer by surprise.

Trailers

With his welder, bulldozer, and an assortment of other tools the farmer went to work. Junkyards, both at home and elsewhere, were scavenged for suitable raw materials. Especially popular were the running gears of forgotten automobiles and trucks. Trailers might consist of nothing more than a pair of wheels, a short axle, and a metal frame.[1] Others are more elaborate and have a platform onto which a great variety of tools can be thrown or pulled. The low-slung, four-wheel implement trailer pictured in Plate XV is one of two or three dozen built on an ad hoc basis by a Peotone machinist for farmers in southern

---

[1]See, for example, the Texaco advertisement entitled "Builds Pickup Carrier for Harrow and Other Machines," *Farm Journal*, October, 1956, p. 36.

Pl. XV.--Though mantled with farmyard refuse in late winter, this four-wheel trailer will be busy toting implements around Will County come the spring.

Will County during the 1940's and 1950's. Farmers would bring in as many parts as they could find, tell the machinist what size trailer they wanted, and perhaps stay around to lend a hand in its construction. The plank floor rests on twin I-beams. Loose planks are inclined up the sides for loading and unloading. The owner of this particular trailer has had several good chances to sell but feels it is far too valuable, both for him and neighboring farmers who use it as much as he does. "It may disappear for days," he said, "but sooner or later somebody will come towing it back."

Docks

Men who depend on flatbed farm trucks and ordinary farm wagons for implement and tractor transport may encounter a loading problem. Truck and trailer beds are frequently too high for an inclined ramp. Even when the truck bed lifts, an extra boost may be necessary. According to a 1963 <u>Farm Journal</u> advertisement, a Nashville, Indiana, farmer handles:

> one of the toughest types of farming--on scattered plots, where equipment has to be moved from one field to another . . . [by easing] the back wheels of his International A-160 [truck] into a ditch, and tilt[ing] the dump bed

to receive and discharge a tractor.[1] Other farmers have built special loading docks to solve the problem. Just a few minutes work with his bulldozer was enough for a young Jackson farmer to create the two-foot-high dirt dock shown in Plate XVI on the bluff overlooking East Locust Creek. Farther upstream at his 100-acre outlier is a duplicate of this one. When the land in the dock area is flatter, a bit more effort is required (see Plate XVII).

Towbars

It is time consuming and risky for a farmer to commute back and forth from an outlying tract to the farmstead on the tractor even if the implement can be left behind. To remedy this, many farmers have installed a homemade (or more recently, commercially made) towing system so the pickup or jeep can be taken to the field at the same time as the tractor and implement. Most prefer to pull the utility vehicle behind the tractor using a towbar similar to the one shown in Plate XVIII. When not needed, it fastens out of the way against the truck's grill. Men who still have tractors with closely set front wheels can mount a towbar between these wheels (Plate XIX) and tow behind the pickup. A Muscatine County, Iowa, farmer does it another way (see Plate XX). He built a sturdy little trailer to carry just the tractor's front wheels. The tractor is driven up the ramp onto the concave trailer platform and chained in place to secure it for towing.[2]

Discouraging Vandalism and Theft

Parked in a farmyard, a tractor is less of a temptation for aberrant adolescents and professional tractor thieves than one parked in a field beside a lonely road. When a familiar farmstead is unavailable, farmers are not ashamed to seek overnight lodging for their equipment in the lot of a stranger who lives near the outlying land. "Farmers are especially understanding when it comes to looking after the property of another farmer," said a Wilton-area man with land ten miles from home. The Florence Township resident who rents the land in Wilton out of which the lot occupied by St. Patrick's Church was carved, pulls in close to this sacred structure when he leaves for home. "The Lord watches

---

[1] Eaton Axle advertisement, Farm Journal, May, 1963, p. 52.

[2] This photograph along with a more elaborate description of the trailer appeared in "Worksavers You Can Use in the Field," Prairie Farmer, February 6, 1971, p. 32.

Pl. XVI.--Simple dirt loading dock in Jackson Township.

Pl. XVII.--A more elaborate dock located at Boynton in northern Sullivan County, Missouri. The weeds on top are growing in the dirt used to fill the space between the sides.

Pl. XVIII.--Pickup truck equipped with towbar. Faintly discernible to the left of the truck are the tracks made by the tractor as it was moved out across previously plowed ground.

over our equipment," remarked his wife. Farther east on the Peotone-Wilmington Blacktop, another Florence Township man rents a tract of Wilton land. After the old house on the 160 acres burned and vandals later damaged his tractor (see Chapter VI), he heard of a family seeking a house-trailer site. For $10 per month and a promise to mow the grass and watch his implements, the farmer permitted the family to park their trailer where the old house had stood (Plate XXI).

Lacking a friendly farmstead or wary watchman, dispersed farmers may have other alternatives. Many try to conceal equipment from the view of travelers by leaving it at the back of the tract or in a convenient gully or woodlot. Remnants of once-extensive Osage orange hedge fences in Wilton, while they last, are adequate for this purpose. Sometimes it is easier for a farmer to immobilize than hide a machine. Since modern tractors are fitted with a key-activated ignition, a logical precaution is to remove the key. At least one farmer, however, hesitates to take the key home for fear of losing it or leaving it in another

Pl. XIX.--Tractor towbar attached to front-wheel assembly. (This photograph as well as Plate XX used with the permission of A. M. Wettach, Mt. Pleasant, Iowa.)

Pl. XX. --A homemade trailer built to hold front wheels of a tricycle-type tractor for towing behind the pickup. When chained up, the ramp at the rear serves also to keep the tractor in place.

Pl. XXI.--House trailer parked at the site of a former Wilton farmhouse as a security measure. Corncrib is still used by man who rents the land.

set of clothing. He believes he has solved the problem by caching the key right on the tractor. Several farmers indicated they go even further by shutting off the fuel supply to the engine, removing the ignition coil, or crippling the hydraulic system so that the implement cannot be lifted off the ground. A few prudent individuals make a practice of checking their tractor and combine tanks for sugar and other foreign substances before starting up the next day.

Gasoline thefts continually plague Wilton farmers and to a lesser degree farmers in the other two study townships. The simplest solution is to use tractors that burn diesel or liquid propane.[1] But even if a man does turn to these alternative fuels, the chances are good he will need gasoline for his combine, farm trucks, and various small engines used around the farm. A buried gasoline storage tank is more costly to install than an exposed overhead model; but

---

[1]For more information on tractor types see: J. A. Weber, Comparing Costs of Operations for Gasoline Tractors, LP-Gas Tractors, and Diesel Tractors, University of Illinois, Cooperative Extension Service Circular 944 (September, 1966).

when the pump is locked, pilferage ceases.[1] Gravity-operated overhead barrels (Plate X) can be locked, too; but one wise Wiltonite deems it cheaper to leave his tank unlocked and let the kids have what they want. Determined gasoline thieves, he knows from experience, "will cut the hose on a locked barrel, take what they can use, and let the rest run out on the ground."

Jackson Township Cattle

Trailing

Although no one knows where or when the practice originated, a number of Jacksonians, especially the veterans of many drives, do a good job of fooling their cattle into thinking the floors of the bridges are sound. Ahead of the trailing herd they scatter straw or even hay on the bridge approach and planking. The cattle are bunched up, so that the scuffling of the leaders will not uncover holes and cracks to frighten the laggards, and pushed through.

Checking

To simplify the chore of looking after cattle at outlying farms, Jackson farmers have been known to use everything from horses to motorcycles to airplanes. With a stock rack on the pickup, a farmer can easily haul his horse several miles, saddle up, and ride off for a quick count of the cows and their calves (Plate XXII). According to one of the Jackson farmers who admitted using his kids' motorcycle for pasture surveillance, "It's not as good as a horse, but you don't have to catch it first." Another innovative farmer claimed he has been accused of plowing with his motorcycle and planting with his light plane. While denying these accusations, he did admit to flying up occasionally from his Sullivan County headquarters to check stock on land he once owned at the southern edge of Jackson. Even better than going yourself, is finding somebody else to count and care for the animals on distant outliers. Such was the fortune of a well-to-do but ailing landowner who in 1969 was renting 250 acres of pasture from a retired farmer 18.6 miles east of his Jackson home on U.S. 136. The rent was reasonable, the lespedeza lush, and the fellow did a good job of watching the stock and reporting any problems.

---

[1] A buried tank, on the other hand, can be a serious burden to a farmer when the electricity supply to the pump is interrupted.

Pl. XXII.--Common sights in Putnam County--pickup with stock rack, farmer's horse, corral, and cattle loading chute.

Branding

Missouri in 1969 did not have a statewide brand registration and enforcement law.[1] Cattlemen were, however, encouraged at the county level to adopt a brand and to register it with the county clerk. As of July 29, 1969, sixty-seven Putnam County parties had availed themselves of this opportunity.[2] The first brand registration occurred prior to March 19, 1943 (date of the second), but no date was given. By 1959, only eight farmers had registered. Thus in the decade prior to July, 1969, the remaining fifty-nine came forward. According to the local veterinarian, the sharp upswing in branding was a result of (1) the rustling problems of the 1960's, (2) the great expansion in farm size and consequent scattering of pasture from one end of the county to the other, and (3) farmer pride in the quality of their animals. One Jackson pasture renter who lives north of Unionville near the Iowa line was able to pin down perfectly his first year on Jackson land (1964) by recalling he had registered his brand in anticipa-

[1] A statewide Missouri brand law went into effect January 1, 1972.

[2] Putnam County, Missouri, Brand Record.

tion of taking his cattle far from home for the first time. A host of other farmers who had not registered their brands were actually marking cattle in one way or another. Most felt an unregistered brand would suffice just as well as an official one if an honest man happened to find the animal. If he were dishonest, it would matter little whether the brand was registered or not since Putnam brands carried no weight outside the county.

What Equipment Manufacturers and Tire Makers Have Done

Implement Portability

Manufacturers of farm implements attacked the farmer's portability dilemma on several fronts. Although a single implement, especially in later years, is apt to incorporate several transport-improvement techniques, we have grouped them into four rather distinct categories for this analysis. The categories are (1) tractor-mounted implements, (2) wheel-mounted implements, (3) narrowing devices, and (4) implement trailers. The first and second serve to separate metal blades and other soil-working apparatus from hard road surfaces. This speeds movement and at the same time reduces tool and road damage. The third makes it possible to have wider working widths in the field without sacrificing too much in the way of mobility. The fourth helps solve the road-contact problem as well as some of the width difficulties.

Tractor-Mounted Implements

After being out of the tractor-building business since 1928, Henry Ford decided in the late thirties to begin again. Now Ford is one of America's major agricultural machinery producers. His decision to retool for tractors was mainly the result of an oral agreement with Harry Ferguson, an Irish engineer, who had perfected the first practical hydraulic hitching system. Ford Motor Company would mass produce a low-priced, light-duty tractor featuring the Ferguson or three-point hitch which, among other things, permitted the farmer by means of a hand-operated lever to lift his tool free of the soil for turning or crossing ditches and free of the road for those hasty trips to the eighty out on the highway.[1] To go along with the three-point hitch, they developed a completely new line of "wheel-less implements."

---

[1]U.S., Department of Agriculture, "The Development of the Tractor," by E. M. Dieffenbach and R. B. Gray, Power to Produce, The Yearbook of Agriculture, 1960 (Washington: Government Printing Office, 1960), pp. 36 and 43-44.

Portability advantages, as it turned out, were never mentioned in Ford's Farm Journal advertising prior to the dissolution of the pact with Ferguson in 1946.[1] By the fall of 1947, however, when Ford Motor Company announced its new Dearborn line of implements, transportability had become a valuable selling point. The farmer was reminded that with a Ford tractor and its hydraulic hitch[2] he could enjoy "safe, easy transport to and from the field"[3] or that the Dearborn LIFT-TYPE "disc is carried instead of dragged"[4] down the road. Other Dearborn implements were lauded as being equally portable. A 1949 ad read in part: "Every farmer who has had the bother of hauling a spring-tooth harrow over the road . . . will appreciate the Dearborn LIFT-TYPE harrow. . . . "[5] Eventually, hydraulic hitching systems similar to that pioneered by Ferguson became "standard or optional equipment on practically all models of tractors."[6] Ford, nonetheless, still claimed to be "America's Largest 'Pickup and Go' Family," offering more than seventy wheelless implements to the farmer in a 1958 sales pitch.[7]

Tractor-mounted implements were fine for light tractors such as the Ford, but problems developed when manufacturers tried to outfit larger tractors with exceptionally heavy tractor-mounted tools like the disk harrow. Clyde, in a 1956 article, wrote, "for large tractors a disk that is wide enough [to utilize a tractor's power] is too heavy to be carried behind a moving tractor."[8] Thus in the 1950's manufacturers turned to the system of transport we are about to discuss.

---

[1] Ibid., p. 44.

[2] After the Ford-Ferguson breakup, "Ford continued to manufacture tractors, making use of the three-point suspension and hydraulic system. Harry Ferguson, Inc., also continued to manufacture tractors with a three-point suspension and hydraulic system, making use of engines from another manufacturer." Ibid.

[3] Ford advertisement, Farm Journal, September, 1947, pp. 70-71.

[4] Ford advertisement, Farm Journal, October, 1947, pp. 86-87.

[5] Ford advertisement, Vedette, April 8, 1949, p. 6.

[6] Power to Produce, pp. 36-37.

[7] Ford advertisement, Farm Journal, April, 1958, pp. 20-21.

[8] A. W. Clyde, "Disk Harrow Design Improvements," Agricultural Engineering 37 (March, 1956): 173.

Wheel-Mounted Implements

If someone had conducted this investigation twenty years earlier, farmers in the Midwest would have been complaining not about implement width but rather about the backbreaking task of manhandling implements on and off trucks and wagons. Not infrequently, farmers' 1969 responses to the present author's questions about portability problems reflected their experience (or lack of experience) with that old, laborious transport system. To a man now in his forties or fifties, the ability to take off and go with a big piece of machinery riding on its own rubber tires often compensates for other impediments to movement that younger farmers find intolerable. "At least we don't have to load them anymore," sighed a middle-aged Deer Creek tenant. "Before getting the new disk," recalled a Wiltonite, "I could never pull out for another place unless the boys were around to help me load the old one."

A number of common farm implements--moldboard plows, rakes, manure spreaders, wagons--were, of course, riding on steel wheels long before pneumatics appeared in the barnyard. After the introduction of rubber tires, farmers were urged "to put Goodrich SILVERTOWNS on implements as well as tractors"[1] or to buy replacement tools already "mounted on pneumatic tires for swift smooth movement on the highway or in the field."[2] There were, on the other hand, several implements, including most of those used for tillage, that either had to be heaved onto a carrier or dragged over the road. Among these wheelless tillage tools we would have found rotary hoes, row-crop cultivators, field cultivators, and various members of the harrow family. None, however, gave the farmer as much trouble as the disk harrow. This is where the portability breakthrough came.

The disk harrow, or disk as it is commonly called, performs before the plow to incorporate crop residue from the previous year and encourage rotting or after the plow as a second step in getting the land ready for a new crop. Disks thrive on weight because this is a major factor in blade penetration. All other things being equal, the greater the weight per blade, the greater is the disk's effectiveness. Today on Corn Belt farms we could find three-ton, four-ton, and larger disk harrows at work. And even in the late 1940's and early 1950's, when our story of the wheel-mounted model begins, the common disk

---

[1] Goodrich advertisement, Farm Journal, September, 1939, p. 30.

[2] New Holland (rake) advertisement, Farm Journal, February, 1949, pp. 98-99.

being handled by farmers must have weighed nearly a ton. Is there any wonder they were complaining?

Not infrequently, an innovation in the agricultural equipment industry has its origin in the shop of an inventor-farmer trying to design something better for his own use. Such a man was Hugh Cooper who, in the late 1940's while farming near Bloomington, Illinois, developed a wheel-mounted disk. In 1950, a shortline,[1] Kewanee Machinery and Conveyor, purchased the rights to the Cooper disk and that fall put four models ranging in width from 7'1" to 14'5" on the market. Each consisted of two gangs of blades, one in front and one behind the pair of pneumatic rubber transport wheels[2] (Plate XXIII). A press release appearing in Farm Implement News noted, "The gangs are raised and lowered by a hydraulic system. With the gangs raised clear of the ground, the implement can be transported without wear on the discs."[3]

Kewanee's wheel-mounted disk was an immediate hit. "We could not make enough to keep up with the demand from the very first year that we introduced it," wrote Glidden from Kewanee.[4] For years after its introduction, Kewanee did little disk advertising; farmers were so eager to buy that it sold itself. Later, at the behest of their dealers who were witnessing a surge of demand for the improved portability of other implements, Kewanee created and successfully marketed the first wheel-mounted mulcher.[5] By 1956, they were also marketing a transportable drag harrow.[6]

Competing firms quickly went to work once the popularity of Mr. Cooper's invention became evident. It took Farm Tools, another shortline, little more than a year to introduce their MOBIL-DISC and begin telling farmers, "IT'S FAST! On the road... in the field... travels at top tractor speed."[7] Major com-

---

[1] Shortline companies offer an incomplete line of farm equipment, and most specialize in implements rather than tractors or combines. Kewanee, for instance, makes several different kinds of tillage tools, tractor blades, wagon (running) gear, grain augers, and grain elevators. Several articles on the shortline appeared in the January 7, 1970, issue of Implement and Tractor.

[2] Letter from D. E. Glidden, Sales Manager for Kewanee Machinery and Conveyor, July, 1973.

[3] "Kewanee Disc Has Hydraulic Control," Farm Implement News, November 25, 1950, p. 38.

[4] Letter from Glidden.

[5] Ibid.

[6] Kewanee advertisement, Farm Journal, February, 1956, p. 53.

[7] Farm Tools advertisement, Farm Journal, March, 1952, p. 82.

Pl. XXIII.--The Kewanee MODEL 10 wheel-mounted disk harrow. Produced in the period, 1950-53, it is just over eleven feet wide. (Photo courtesy of Kewanee Machinery and Conveyor--A Division of Chromalloy American Corporation.)

panies such as Deere and International Harvester, which had accounted for virtually all the pre-Kewanee disk business, soon followed the shortliners' lead with their own wheel-mounted offerings.[1] Even Ford made a bid for a share of the wheel disk market.[2] By 1973, of Deere and Company's fourteen basic disk models, twelve were mounted on wheels, one sold either with or without wheels, and one was tractor-mounted.[3] Each of Kewanee's seven 1973 models was equipped, like the first, with the capability of raising up out of the ground on its own wheels and following the tractor out the lane and down the road.[4]

We should mention that wheels on a disk harrow serve other functions in addition to the primary one just discussed. When the disk is working, the farmer can hydraulically change the wheel setting and thereby determine how deeply the blades penetrate. For the sake of seedbed uniformity, it is sometimes better to avoid the greatest possible penetration. Transport wheels can also be dropped down to raise the blades completely free of the soil on turns and when crossing streams, grass waterways, or similar areas on the farm that he wishes not to decimate. Size of the wheel-mounted disk is limited only by the horsepower of the tractor and width restrictions facing the farmer when he attempts to move it.

Narrowing Devices

Before the advent of big wheel-mounted tools like the disk harrow, midwestern farmers had little problem with width on the road. In the first half of our study period there appears only one reference to width reduction on the pages of the Farm Journal and that was to call farmers' attention to a spike-tooth harrow capable of folding for passage through gates.[5] From the mid-

---

[1] Letter from Glidden.

[2] "You can quickly hook a Ford wheel type [disk] harrow behind any tractor, raise and lock the gangs a full 11 inches off the ground and travel to the field. There's no more lifting and loading... no grinding away of blades on gravel roads." Ford advertisement, Farm Journal, September, 1959, pp. 62-63.

[3] Drawn and Integral Disk Harrows (Moline, Illinois: Deere and Company, 1973).

[4] Many New Products from Kewanee (Kewanee, Illinois: Kewanee Machinery and Conveyor Company, December, 1972), pp. 2-13.

[5] "Four-section harrows fold to 11 feet for gate clearance." Frazer advertisement, Farm Journal, September, 1946, p. 82.

1950's on, however, as manufacturers sought to keep pace with higher tractor horsepower by increasing field widths of implements, many farmers began to encounter the impediments to movement we discussed in Chapter VI. The first Farm Journal advertisement calling attention to the road-width dilemma ran in the February, 1956, issue.[1] Since then, the narrowing devices adopted to help farmers overcome this problem have been a regular ingredient of implement advertising.

It is extremely difficult to generalize about critical transport width. How great must the working width become before the manufacturer takes steps to make the implement more compact for the road? The answer depends on several things. What kind of implement are we talking about? Does it travel on its own wheels, or is it totally dependent on the tractor's hitch for support? Where is it to be sold? Will the transport feature price it out of the market? At the moment, for several midwestern shortlines, the critical width is between thirteen and seventeen feet. Somewhere in this range the switch is made from four-row to six-row equipment. Kewanee's widest inflexible disk requires 15'4" in lateral clearance. Its smallest flexible model, on the other hand, has a working width of 13'5" compared to 10'1" on the road.[2] Noble's largest nonfolding, tractor-mounted field conditioner works at fifteen feet while its smallest folding model of the same tool works at thirteen feet.[3] A Wisconsin firm, Brillion Iron Works, designs its tillage products with the following in mind: "When overall widths extend beyond 16 feet, farmers encounter problems of getting through gates, down narrow lanes or traveling on country roads."[4]

Folding

Whenever feasible, this is the preferred narrowing technique. The implement remains hitched to the tractor just as in the field. Another technique, to be discussed momentarily, requires an alteration in this implement-tractor relationship. Folding is accomplished either by elevating the implement's lateral extremities and perhaps resting them upon a fixed central section or by

---

[1] Kewanee advertisement, Farm Journal, February, 1956, p. 53.

[2] Many New Products from Kewanee, pp. 2-13.

[3] 3-Point Mounted Field Conditioner (Sac City, Iowa: Royal Industries, Noble Division, n.d.).

[4] Brillion Pulvi-Mulcher (Brillion, Wisconsin: Brillion Iron Works, n.d.), p. 2.

moving the extremities in a horizontal plane to a transport position in front of or behind their field position.

The Up-and-In Fold.--Most disk manufacturers use this method (Plates XXIV, XXV, XXVI). By folding the wings up and in they add weight to the main gangs which can then function as "a narrower, deeper-working disk" when field conditions are demanding.[1] In some models the wings flex vertically while at work, thus permitting them to follow the contour of the land better. The folded disk is also more compact in storage. These added advantages of the up-and-in fold, notwithstanding, greater portability is really what the manufacturer is seeking. One of the most portability-conscious of all implement makers is a Kansas firm, American Products, Incorporated. They label their CRUST BUSTER folding offset disc "the down-the-road disc...for the down-the-road farmer" and go on to remind potential customers:

> If you're an expanding farmer, chances are you're not lucky enough to expand next door. Your land may be scattered all over the county. No matter how good a disc is in the field, it's of little value unless you can get it to the field when you need it. Unless a tool is as portable as your tractor, it doesn't get the CRUST BUSTER name.[2]

Field cultivators and other seedbed tools commonly employ the up-and-in fold, too. Noble's largest tractor-mounted soil conditioner is shown in Plate XXVII working down a twenty-foot swath of Iowa ground. In Plate XXVIII the outriggers have been hydraulically raised and locked in place over the center section for transport. Deere reduces the width of its 42.5-foot field cultivator to 19.5 feet by hydraulically lifting the wings to the vertical position (see Plate XXIX). If the farmer is willing to spend extra time cranking up the outriggers by hand, this type of mechanism is also available (see Plate XXX).

An exceptionally difficult piece of equipment to narrow is the row-crop planter (Plate XXXI). Even with its longest wheel-mounted planter (twenty-four feet), Deere and Company offers the farmer no alternative except to pull it on the road at that width.[3] For one thing, planters are extremely delicate tools. For another, folding means tipping the fertilizer, pesticide, and seed containers up so that spillage will occur unless they have been emptied or meticulously

---

[1] Wing-Type Wheel Tandem Disk Harrows (Yazoo City, Mississippi: AMCO--A Division of Dynamics Corporation of America, n.d.).

[2] CRUST BUSTER Folding Offset Discs (Spearville, Kansas: American Products, Incorporated, 1972).

[3] Pull-Type and Mounted Planters (Moline: Deere and Company, 1972), p. 4.

Pl. XXIV.--The MODEL 1000, largest disk in the current Kewanee line. Shown here ready for transport, the folded width is thirteen feet, or two feet wider than the MODEL 10 (Plate XXIII). The MODEL 1000 will work at 21'2". (Photo courtesy of Kewanee.)

Pl. XXV.--AMCO's huge wheel-mounted tandem disk rests briefly in a Mississippi field. (This photo and Plate XXVI courtesy of AMCO--A Division of Dynamics Corporation of America.)

Pl. XXVI.--Same disk as in Plate XXV shown in the transport mode on a dusty Delta road.

sealed beforehand. Nevertheless, if the farmer is determined to have an easier planter to transport, the Paul Abbott Company in Blytheville, Arkansas, will custom build one for him using their FOLDING IMPLEMENT CARRIER and parts from new or used four-row planters (Plate XXXII). The finished product enables "the man going places" to "cross bridges, travel highways, [and] pass gates. . . ."[1] He is cautioned, however, to remove the contents of the outside containers before folding up for a move.

The Horizontal Fold.--There are several types of horizontal folding patterns. When the row-crop cultivator is mounted near the front of the tractor (usually just behind the front wheels), the sections on either side can be swung forward and locked together. The farmer is, in effect, obliged to push the cultivator ahead of the tractor when on the move. With some rear-mounted tools, the wings are drawn forward toward the tractor and secured. Finally, in the case of unusually wide implements, such as a drag harrow or the Noble CULTI-MATIC field conditioner, it is deemed best to fold toward the rear. In the field, the CULTI-MATIC shown in Plate XXXIII is more than thirty feet wide. From the seat of the tractor, the operator raises the teeth out of the ground, moves the tractor forward until the wings are parallel, and heads for the next job pulling a long load but one which is only 10'6" wide[2] (Plate XXXIV).

The New TRI/FOLD.--Portable Elevator has recently introduced a row-crop cultivator in its Glencoe series which uses both the up-and-in and the horizontal fold to reduce width. The cultivator, which may stretch out to thirty-one feet in the field, is mounted on the tractor's three-point hitch. First, the wings are folded up and in. Then, the whole implement "is shifted forward on the tractor adding front end stability to the tractor in transport"[3] (Plate XXXV). This is a brand new solution to the problem that manufacturers have been fighting for years when they tried to tractor-mount tools that were too heavy for the tractor to carry safely at high speeds. According to a company spokesman,

---

[1] Folding Implement Carrier (Blytheville: Paul Abbott Company, n.d.), p. 1.

[2] Culti-matic Field Conditioner (Sac City, Iowa: Royal Industries, Noble Division, n.d.).

[3] Glencoe Row Crop Cultivators (Bloomington, Illinois: Portable Elevator --A Division of Dynamics Corporation of America, 1972).

Pl. XXVII.--A tractor-mounted Noble soil conditioner at work. Notice the slow-moving-vehicle emblem on the rear of the tractor. (This photo and Plates XXVIII, XXXIII, and XXXIV courtesy of Noble--A Division of Royal Industries.)

Pl. XXVIII. --The Noble Soil Conditioner folded up and in for transport. The small pneumatic tires on the implement serve only in the field.

Pl. XXIX.--Transport position of Deere's 42.5-foot, wheel-mounted field cultivator. (Photo courtesy of Deere and Company.)

Pl. XXX.--A McLean County, Illinois, farm boy cranks up the outriggers on his dad's field cultivator after finishing the tract behind him.

Portable Elevator hopes to incorporate the TRI/FOLD design on other Glencoe tillage tools because of the success enjoyed thus far with the row-crop cultivator.[1]

Altering the Implement-Tractor Relationship

For some tools the relationship change is rather slight; but it does, nevertheless, reduce the farmer's transport width. It is used with implements, such as the hay baler, pull-type combine, or corn husker, which are stationed to the side of the tractor in the field. Either the "tongue pivots to provide correct tractor position for transport," as with the Deere husker,[2] or the implement must be unhitched and the tongue set over manually so it will ride behind the tractor. Even while working, small moldboard plows tend to stay directly

---

[1]Thayer, interview, Bloomington, Illinois, July, 1973.

[2]Corn Pickers and Husker (Moline: Deere and Company, 1971), p. 3.

Pl. XXXI.--This six-row planter takes nearly half the oncoming lane as the farmer swings around a highway curve north of Kokomo, Indiana.

Pl. XXXII.--An eight-row Abbott-modified planter meets a pickup on a Delta road near Blytheville, Arkansas. (Photo courtesy of Paul Abbott Company.)

behind the tractor. But when the number of bottoms on the plow exceeds four or five, the tail will begin to protrude past a line drawn rearward from the tractor's left rear tire. To alleviate this problem, some plowmakers install a device that enables the farmer to reduce his road width with a manual adjustment of the hitch.[1]

Highway towing of other implements including rotary hoes, rolling clod pulverizers, and some of the big tractor-mounted (rear) row-crop cultivators may require a more radical revision of their position vis-à-vis the tractor. To help the farmer move these cumbersome tools, Brillion, Deere, Portable Elevator, and no doubt other firms sell what is variously called an "endways transport attachment" or "lengthwise transport kit" which bolts to and becomes a semipermanent part of the implement itself. The Portable Elevator version is shown mounted on an eight-row (row-crop) cultivator in Plate XXXVI. The kit has three basic components: (1) a superstructural torsion bar running the length of the tool bar and attached to it in several places, (2) a pair of pneumatic tires near one end, and (3) a special hitch and jack assembly at the other end. When the farmer finishes at one place, he hydraulically raises the shovels free from the soil, flips the transport wheels down from the out-of-the-way location they occupied in the field, and cranks down the jack to support the hitch end. Since the cultivator is now standing on its own wheels and the jack, he can unhook the tractor from the regular hitch, pull around to the special transport hitch, and hook up again. Finally, he disengages the jack and drives away trailing a load measuring less than eight feet in width.[2] Deere's thirty-one-foot rotary hoe is being moved with the aid of a transport attachment in Plate XXXVII.

Commercially Built Implement Trailers

For long trips or to expedite the movement of machines that are not equipped for road travel, some farmers still depend on the all-purpose trailer. A young Deer Creek operator uses the tilting cart shown in Plate XXXVIII to haul such immobile tools as the roller seen standing to the left of it. Winched by hand onto the back of the cart, the roller's weight will eventually level it out. Lighter tools like the harrow are thrown on by hand. Fortunately, the disk which is barely visible behind the roller is mounted on its own wheels. The

---

[1] Moldboard and Disk Plows (Moline: Deere and Company, 1972), p. 9.

[2] Glencoe Row Crop Cultivator Transport (Bloomington: Portable Elevator, 1972).

Pl. XXXIII. --Noble CULTI-MATIC field conditioner at work

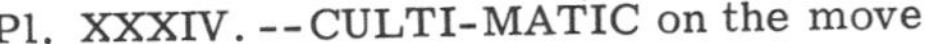

Pl. XXXIV.--CULTI-MATIC on the move

Pl. XXXV.--Portable Elevator's TRI/FOLD row-crop cultivator with weight shifted forward to help keep the tractor's front end down while on the move. (This photo and Plate XXXVI courtesy of Portable Elevator--A Division of Dynamics Corporation of America.)

loaded cart is pulled behind the pickup to the field by the farmer's father or wife while he takes the tractor and another piece of equipment.

Donahue Manufacturing, a Kansas firm specializing in farm carriers, offers an eleven-model lineup of heavy-duty implement trailers ranging in width from five feet to nine feet and in length from twenty-one feet to forty-eight feet. Farmers find the Donahue trailers especially handy, because they are designed to lie flat on the ground during loading and unloading. A tractor can tow an implement onto them without the need for special ramps or tiltable bed. Once the implement is in place, the driver hitches to the trailer's towbar, returns the trailer wheels to their customary road position, and departs.[1]

---

[1] Donahue Implement Carrier (Durham, Kansas: Donahue Manufacturing, n.d.).

Pl. XXXVI. --Portable Elevator lengthwise transport attachment prepared to relocate a row-crop cultivator.

Pl. XXXVII. --A Deere rotary hoe and its transport kit follow the tractor across a railroad track. (Photo courtesy of Deere and Company.)

Pl. XXXVIII.--View of a Deer Creek farmyard. On the right is a tilting implement cart and on the left a roller which depends on the cart for transportation.

A Final Note on Portability

The portability aids we have been discussing naturally increase the cost of a farmer's machinery inventory. Sometimes the exact amount is concealed because the improvement is now standard equipment. Brune estimates that 12 percent of the cost of Deere's medium-duty disk is attributable to the folding mechanism. They are now "introducing a new line of disks where transport considerations involve approximately a fifth of the cost."[1] With other aids it is not so difficult to price portability. Deere offers its transport kit for $340 plus freight from the factory.[2] Portable Elevator sells its straight eight-row cultivator for just over $1100. For $270 more, the farmer can obtain it with the lengthwise transport attachment at the time of purchase. Later on, the same attachment would cost him $325.[3] Abbott will sell just the basic toolbar so that cus-

---

[1]Letter from Rey Brune, Public Relations Officer for Deere and Company, August, 1973.

[2]Ibid.

[3]Telephone conversation with Jerry Nussbaum, Sales Manager for Portable Elevator, August, 1973.

tomers (if they wish) can design their own implement. A 21'6" nonfolding bar lists for $300 compared to $530 for the folding model of the same size.[1] Donahue's implement trailers range in price from $826 for the smallest to $2263 for the largest.[2]

Tire Durability

In 1962, after years of paying lip service to excessive road wear, Goodyear and Firestone, the two major tractor tire makers, finally undertook a serious effort to counteract it. We intend, in the paragraphs that follow, to trace the evolution of tread design from the late 1930's through the 1960's. We shall note in particular the awakening of interest in road-wear resistance in the 1940's, the feeble attempts by the tire makers to convince farmers their designs in the forties and fifties were superior to those of competitors, and finally the encouraging developments of the last decade.

Open Center or Closed Center?

Earliest pneumatic tractor tires were available in two relatively shallow tread patterns--a closely spaced, nondirectional knob and a slightly wider spaced but still nondirectional bar.[3] Neither pulled in sticky mud very well because the mud tended to pack the spaces, leaving a virtually smooth surface to contact the soil. Firestone, in the mid-1930's, first began using a connected- or closed-bar directional design in which the bars were set at a 45-degree angle to an imaginary line drawn across the tire parallel to the axle[4] (see Figure 14). Goodyear announced its version of the 45-degree directional bar in September, 1937.[5] But instead of connecting the bars at the center like Firestone, they chose to leave a continuous open space. Thus by 1939 the stage was set for a decade of conflicting claims.

---

[1] Folding Implement Carrier.

[2] "Suggested Retail Prices-Donahue Farm Implement Carriers," Effective March 15, 1973.

[3] P. J. Forrest, "Effects of Improper Inflation Pressures on Farm Tractor Tires," Agricultural Engineering 35 (December, 1954): 833.

[4] For a recent discussion of bar angle and other tire characteristics see: Melvin E. Long, "Turning Engine HP into Drawbar HP: Choosing the Right Tire," Implement and Tractor, August 7, 1969, pp. 28-30.

[5] Goodyear advertisement, Farm Journal, September, 1937, pp. 32-33.

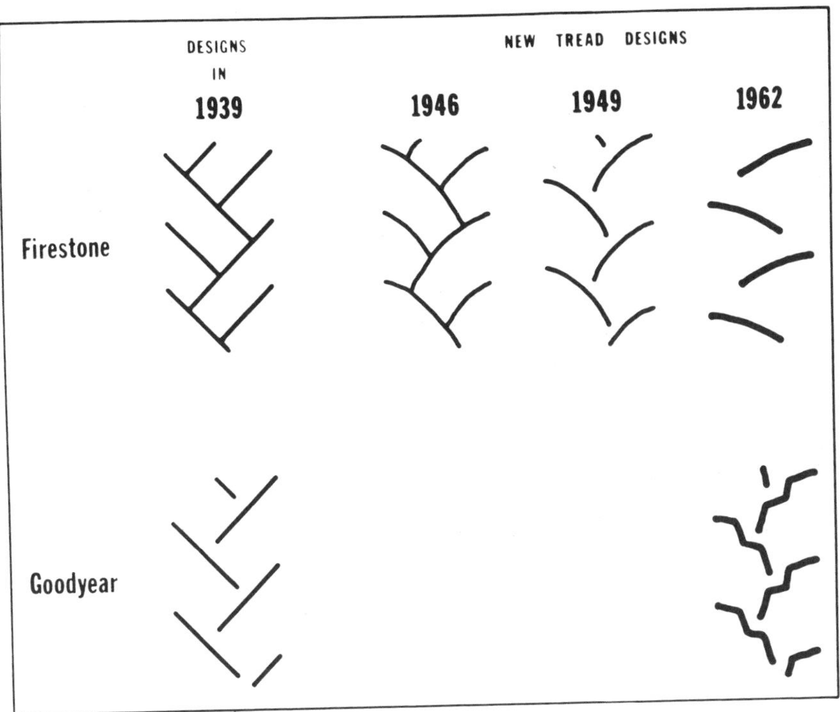

Fig. 14.--Rear Tractor Tire Tread Designs of Firestone and Goodyear: 1939-1962.

Since roadability has always helped sell pneumatic tractor tires, we are not surprised to find early references to the road in Farm Journal tire advertisements. Goodyear, for instance, was reminding farmers in a 1939 publicity statement that their "lug bars overlap evenly at [the] center, giving continuous support on hard roads."[1] Firestone quickly countered by noting their product featured "continuous triple-braced traction bars" joined together for smoother riding.[2] The first actual reference to road wear probably came late in 1941, but we cannot be entirely certain. The Firestone advertisement features a drawing of a tractor rolling on a hard surface that appears to be a road. We are told,

---

[1]Goodyear advertisement, Farm Journal, January, 1939, pp. 30-31.

[2]Firestone advertisement, Farm Journal, March, 1939, p. 8.

"There are no unsupported bars which wobble and wipe, and lose their sharp, biting edges."[1]

During the war, tire makers had all the business they needed. Advertisements stressed the efforts being made to develop synthetics and to supply the armed forces with tires.[2] Of some significance to our current topic was a 1943 U.S. Rubber Company ad reminding farmers that the Office of Price Administration was prepared to authorize tractor tire purchases for the purpose of converting from steel to rubber if it was "necessary to operate tractors on highways between farms."[3] As the war ended, the tire companies resumed their own skirmishing. Goodyear revealed to Farm Journal subscribers the favorable (to Goodyear) results of some actual field tests matching open-center against closed-center treads.[4] Goodrich was also in an objective mood two years later when it polled farmers for their opinions on tread design. Of every three questioned, two preferred the open centers made by Goodrich and, of course, by Goodyear, too.[5]

Curved Lugs or Straight?

The first tread change by a major company since the thirties and the first definite comment on road abrasion came late in 1946. Firestone in a two-page color announcement abandoned straight bars for a closed-center, curved design. Praising this "Revolutionary New Tractor Tire," Firestone contended their traction bars would not "bend, break nor push through the cord body. And on the highway they don't bounce, wobble nor wear like unsupported bars on an open center tire."[6] Both U.S. Rubber and Goodyear answered Firestone in June of 1947. The U.S. ROYAL tire, with a partial connecting rib down the midline of the tread could, according to their advertising agency, help farmers beat "costly road wear" by guarding "against destructive lug-bending and

---

[1] Firestone advertisement, Farm Journal, December, 1941, p. 10.

[2] Especially 1942 and early 1943.

[3] U.S. Rubber advertisement, Farm Journal, June, 1943, p. 69.

[4] Goodyear advertisement, Farm Journal, December, 1944, p. 3.

[5] Goodrich advertisement, Farm Journal, September, 1946, p. 20.

[6] Firestone advertisement, Farm Journal, November, 1946, pp. 46-47.

scuffing."[1] Goodyear asked farmers to recall their ten years of experience with the straight-lug SURE-GRIP noting, they "always have the same number of lugs on the road. This eliminates pavement 'stubbing' and distributes the wear equally."[2]

The longstanding feud between open and closed centers finally fizzled out in the early fifties. Firestone had announced a "Sensational New Open Center Curved Bar Traction Tire" in December of 1949 to accompany its closed-center CHAMPION.[3] In a later issue they explained, "Some farmers prefer open center tires, while others demand Traction [closed] Center."[4] By late 1951, however, Firestone was trumpeting just the open-center version.[5] The only feature, on the surface at least, that distinguished Firestone from Goodyear during the 1950's was the fact that one sold tires with curved lugs while the tires made by its major competitor had "ruler-straight lugs."

Goodyear continued to praise the virtues of its venerable tread design while simultaneously noting how the lugs had been deepened, especially at the tire shoulders, so that "It rolls smoothly on the road."[6] Carcass cord was also improved "for longer life, for better relugging."[7] In other words, the lugs on the Goodyear tires molded in the fifties were deeper than before, but if they were to wear out, the stronger cord would enable the farmer to retread and continue using the same carcass. Through the middle and late 1950's advertisers for tire companies stressed tread rubber compounds, amount of rubber at critical spots, cord strength, and tire loaning policies to accommodate the farmer while his was being repaired. Silence on the matter of roadability was an excellent indication that farmers and tire makers alike were dissatisfied with excessive road wear. Goodyear bought a two-page Farm Journal spread in 1958 to announce the "World's Newest, Workin'est Traction Tire," but few obvious

---

[1] U.S. Rubber advertisement, Farm Journal, June, 1947, p. 35. "Bar" and "lug" are here synonymous.

[2] Goodyear advertisement, Farm Journal, June, 1947, p. 3.

[3] Firestone advertisement, Farm Journal, December, 1949, pp. 24-25.

[4] Firestone advertisement, Farm Journal, March, 1950, p. 57.

[5] Firestone advertisement, Farm Journal, December, 1951, p. 25.

[6] Goodyear advertisement, Farm Journal, March, 1954, p. 3.

[7] Ibid.

changes had been made.[1] Everyone associated with the tractor tire seemed to be waiting for something to happen.

Bar Shape and Angle?

Within a few months of one another in 1962, Firestone and Goodyear introduced their contemporary tread designs.[2] For Goodyear, it marked the first real tire-footprint change in a quarter century. Firestone, for its contribution, halved the conventional 45-degree bar angle thus setting a precedent that virtually all tire makers would eventually follow.

Goodyear did not immediately phase out its straight-lug TRACTION SURE-GRIP. Instead, the new SUPER-TORQUE was offered as a premium tire for those with big tractors who needed extra traction in the field and better road mileage than had been delivered by previous tires. Retaining the essence of a 45-degree bar, Goodyear added several kinks or angles to it. This put more rubber in contact with the ground and braced the lug to discourage folding under while on pavement. By May, 1962, Goodyear was telling farmers the SUPER-TORQUE "costs you nothing... if it doesn't outpull and outwear any other tractor tire in America."[3] Eventually, Goodyear began offering this "angle-braced" or "zig-zag" tread on its cheaper lines of agricultural tires, because the SUPER-TORQUE was too expensive to suit most potential replacement tire buyers.[4] In addition, without making any special note of it to Farm Journal readers, they had by 1969 reduced the basic bar angle to around 30 degrees.

FIELD & ROAD was the appropriate name chosen in 1962 by Firestone for its new tractor tire. Farmers were told they would "outpull any replacement tire" mainly because the bars were spaced more widely and set at an unheard-of 23-degree angle.[5] In the introductory statement, Firestone did not promise their tires would outwear all other competitors on the market. They did, however, warn farmers not to be surprised if that turned out to be the case. And sure enough, through the sixties, Firestone's advertisements were replete

---

[1] Goodyear advertisement, Farm Journal, March, 1958, pp. 14-15.

[2] Goodyear advertisement, Farm Journal, February, 1962, pp. 5-8. This four-page spread was the largest bought by a tire company in the three-decade period. Firestone advertisement, Farm Journal, May, 1962, pp. 50-51.

[3] Goodyear advertisement, Farm Journal, May, 1962, p. 3.

[4] Goodyear advertisement, Farm Journal, March, 1968, pp. 22-23.

[5] Firestone advertisement, Farm Journal, May, 1962, pp. 50-51.

with testimonials from satisfied FIELD & ROAD owners. Here are a few:

> These tires are the answer for highway travel.[1]
>
> We do a lot of roading and FIELD & ROAD tires wear much less.[2]
>
> Before I got them I wore out two sets of tires a year on the highway pulling a heavy crop sprayer. These tires have over three thousand miles on now.[3]
>
> 1,200 miles of road work first year. Looks like they'll outwear 3 sets of regular tires.[4]

But Still Farmers Complain

The new generation of road-resistant tires had been on the market seven years when the author began soliciting information for this study; yet, complaints about tire wear persisted. Since no attempt was made to learn the actual types of tires being used, it is impossible to say for sure what proportion of farmers had tried the FIELD & ROAD or the Goodyear lineup headed by the SUPER-TORQUE. No doubt some men were complaining about road abrasion when a road-resistant tire would have helped them. On the other hand, it is difficult to see how very many could have avoided contact with these tires because they eventually came as original equipment on new tractors and combines.

It is this author's belief that Goodyear, Firestone, and later Goodrich and the other tire makers have, after finally confronting the road-wear problem, taken sizable strides toward counteracting it. But given more road travel, heavier loads on tires, and a growing number of blacktop miles, lug wiping is apt to keep pace with tire improvements. And so long as it does, farmers will more than likely continue to complain. Premature tire deterioration is simply an expense of noncontiguity.

Assistance from the Public Sector

At the Local Level

Frustrated by narrow country bridges, midwestern farm operators have sought and found sympathetic ears among local road and bridge authorities. Recall, for instance, the cooperative attitude of the Wilton road commissioner

---

[1] Firestone advertisement, Farm Journal, November, 1962, p. 13.

[2] Firestone advertisement, Farm Journal, February, 1963, p. 6.

[3] Firestone advertisement, Farm Journal, March, 1963, p. 8.

[4] Firestone advertisement, Farm Journal, May, 1965, p. 23.

when a constituent wished to revamp the side of a small, troublesome bridge under his jurisdiction. Cass County is, according to Highway Department official Donald Lake, "tearing out old concrete culverts as fast as possible and replacing them with tubes."[1] More often than not, the tube also extends under the road shoulders, so that there is less need for guard rails. Old metal bridges over Deer Creek are gradually being phased out in favor of concrete structures.[2] Will County is not as ambitious about replacing bridges as Cass, but prestressed concrete models are becoming more common. Although the new prestressed bridge shown in Plate XL was ordered specifically to accommodate farm machinery up to nineteen feet in width, most replacement bridges in Will (and Cass, too) are authorized with auto and truck traffic in mind. Along the border with Kankakee County stand several extremely narrow metal bridges that Will County intends to help replace once its neighbor to the south catches up on bridge-widening projects elsewhere.[3]

For farmers in Jackson and vicinity, the bridge prognosis is less encouraging. Neither the townships nor the counties have funds really to do as the farmers who travel the roads with their machinery would like.[4] Most bridge decisions in northern Missouri are made at the county level. Few new bridges, other than small wooden structures, have been built by the Putnam County Court[5] in the last three decades; but the judges are as understanding as their modest bridge budget will permit. Tubes are being purchased and installed wherever drainage and soil conditions are favorable. In the case of creeks like the Lo-

---

[1]Interview with Donald Lake, Assistant Superintendent of Highways, Cass County, Indiana, July, 1970. A tube is a rigid, corrugated, metal drainageway.

[2]The westernmost bridge over Deer Creek in Deer Creek Township was scheduled for replacement in 1970. "State Board Approves New Bridge in County," Pharos-Tribune, August 3, 1970, p. 1 (see Plate XXXIX).

[3]Interview with Ralph Yunker, Chief Engineer, Will County Highway Department, Will County, Illinois, March, 1970.

[4]Jackson was cited by a local tube salesman as one of the most miserly townships in his territory when it came to levying taxes for bridge work.

[5]Administrative affairs are handled in the great majority of Missouri's counties by a three-member elected court. Although called judges, the three actually have no judicial functions. A presiding judge is elected by all the county's residents and the other two by residents of either the northern and southern districts or eastern and western districts depending on the shape of the county. Jackson Township is in the western district of Putnam County. Harry S. Truman served as a judge of the Jackson County (Kansas City) Court (1922-4) and as its presiding judge from 1926 until elected to the United States Senate in 1934.

Pl. XXXIX.--This metal arch bridge across Deer Creek three miles north of Young America is unsafe for heavy traffic and will be scrapped.

Pl. XL.--A new prestressed concrete bridge in Wilton Township.

custs where only bona fide bridges will do, the court sometimes agrees to improve circulation by widening an existing structure. Thus when East Locust in central Jackson was straightened recently, the steel from an abandoned eleven-foot bridge was moved to the new crossing point and the width of the roadway increased to more than thirteen feet. As one farmer put it: "The county is doing its darndest to see that four-row equipment can move on the back roads." To encourage even more widening, some farmers have offered to help pay for it. The court in 1969 was listening with interest to these offers.

At the State Level

If a farmer has his implements of husbandry properly flagged, signed, and lighted[1] and if he drives prudently, Missouri, Illinois, and Indiana allow him to do just about as he pleases when it comes to machinery movement. None of the three states requires him to register or license his farm tractors and combines if they are moving from field to field or to and from places of repair or delivery.[2] As far as these states are concerned, anyone may operate a tractor on the road in the situations just described even without a valid driver's license.[3] Neither age nor a suspended license is deemed sufficient reason to deny a person the privilege of driving a farm vehicle. Farm equipment is specifically exempted from the need to comply with size regulations.[4] Missouri statutes, for instance, limit the road width of vehicles to 8 feet, height to 13.5 feet, and length to 40 feet but go on to say, "these restrictions shall not apply to agricultural implements operating occasionally on the highways for short distances."[5] To the Missouri highway patrolman responsible for Putnam County in 1969, "a farmer's tractor and implement could take up the entire pavement and

---

[1]Illinois demands a red flag and amber flasher, Indiana a slow-moving-vehicle emblem (SMV) and a red or amber flasher, and Missouri the SMV.

[2]Illinois, Revised Statutes, Annotated (Smith-Hurd), 3-103; Indiana, Revised Statutes, Annotated (Burns), 47-2601; Missouri, Revised Statutes, Annotated (Vernon), 304.260.

[3]Illinois, Revised Statutes, 6-102; Indiana, Revised Statutes, 47-2702; Missouri, Revised Statutes, 302.080.

[4]Illinois, Revised Statutes, 15-102; Indiana, Revised Statutes, 47-530; Missouri, Revised Statutes, 304.170.

[5]Missouri, Revised Statutes, 304.170.

still be legal."[1] Illinois law does specify, however, that its agricultural exemption applies only "during the period from sunrise to sunset."[2]

---

[1]Interview with Kenneth Swon, Missouri Highway Patrolman, July, 1969.

[2]Illinois, Revised Statutes, 15-102.

CHAPTER VIII

SUMMARY OF FINDINGS

In keeping with the title of the present study, Farmers on the Road, specific questions concerned with (1) interfarm migration and (2) the farming of noncontiguous land have been pursued in the preceding chapters, following their listing at the close of Chapter I. They have served as guides for an investigation conducted in three study areas, each a township in the American Midwest. At this point, it is possible to present answers to these questions in a conveniently assembled form.

Interfarm Migration

Residentially, farmers are far more stable today than they were in the early 1940's. Short-distance moves continue to be the rule for most of those who choose to relocate. Young farmers have tended to be more mobile than their older neighbors. Newcomers who previously had been living in the same county as their new farms greatly outnumbered those whose journeys brought them across a county boundary. Little evidence was found of topographic effects on migration patterns, but a distinct cultural boundary across one township did seem to have repelled movers.

Farmers moved for many different reasons. More land or better land were strong inducements. Even stronger was outright ouster of tenants by their landlords, private and corporate, or of tenants and owner-operators by the federal government, which in one area desired a large block of farmland for its own use. A large number of persons left good farms to escape what family members considered intolerable conditions at their farmsteads or because of the mud on the roads leading to them.

When the game of musical farms was an accepted part of the farming experience for thousands of midwestern agriculturalists, nobody gave it a second thought. Then and later, in searching for the next farm, families have depended a great deal on personal contacts with owners of farmland. Friends, relatives, and business acquaintances have served regularly as valuable information intermediaries. Availability notices in the local newspapers and signs at the front gate seldom have been employed to advertise. Farm periodicals have attracted advertisements for expensive and specialized farms that are hard to rent or sell.

March 1 has been the traditional rural moving day in the Midwest. From the standpoint of the farm year it is an excellent choice. Crops can be planted and young livestock dropped at the new place later in the spring. Hay, which has been stored for the winter, will by the end of February have been pretty well exhausted, so that the farmer does not have to move it. On the other hand, for getting in and out over muddy roads with heavily loaded trucks, early March was (and in some places, still is) a terrible time to be moving. To many concerned observers of rural lifestyles, the first of March came to symbolize the frustration of failure at one place and the hope for success at the next.

The Farming of Noncontiguous Land

Nonfarmstead land is much more common now than it was before World War II. A rental agreement is the most popular means of acquiring it. With farmstead tracts, on the other hand, purchase and inheritance are more important. Although many farmers fear or dislike the arrangement, demands by private landlords for cash rent--as against a share of the crop--on their nonfarmstead land are becoming widespread. Several thousand acres of public land are cash-rented annually from the federal government by farmers living in and around one of the townships studied.

A growing proportion of nonfarmstead land is not contiguous to the farmstead tract. Distance between the farmstead and an outlier can affect a farmer's decision whether to acquire it or not. The distance farmers are willing to travel was not predictable from tract size when we considered all tracts but was when we aggregated tracts of nearly equivalent size. Forty-acre tracts, for instance, do not lure farmers as far from home as do eighty-acre tracts. Land quality proved difficult to gauge; but by measuring it indirectly (through land use practices), we discovered a tendency for farmers to travel farther for the better land. When a tract lay many miles from its operator's farmstead, the chances were good that there was some special tie between it (or its owner) and the farmer. Young farmers tend to go farther for land than do their elders. An exception was the 50-59 age cohort, which in one study area actually did the most travelling per 1,000 acres of land.

Noncontiguous land is more expensive to operate than land in or contiguous to the farmstead tract. Time, fuel, and tire tread are wasted, not to mention the risk of damage to delicate parts of a machine, such as a planter, when it must be moved rapidly from tract to tract over rough roads. Estimates of time lost and fuel burned on the road were made for representative 1969 farmers.

Country roads are seldom conducive to easy machine movement. Narrow shoulders, signs on the shoulder, and narrow bridges can inhibit or even prohibit circulation with wide equipment. Travel by tractor or self-propelled combine on roads carrying high-speed automobile and truck traffic is hazardous for all parties but especially for the farmer. Even if an accident does not result in injury or death to him, he faces the prospect of a financial setback due to a lawsuit or the inability to get the damaged machinery repaired quickly. Livestock can be moved to outlying parcels by truck, but many farmers still use the over-the-road method of their forefathers. Several drovers are needed to push even a modest herd of cattle, and then some are likely to stray from the road _en route_. The dispersed farmer also has to contend with inadequate knowledge about the daily condition of the soil and livestock at his outliers. Threat of animal, stored-grain, and stored-equipment thefts is increased by noncontiguity.

The burdens of noncontiguity have been eased somewhat through actions taken by the farmers themselves. Some have tried to concentrate their holdings in more easily accessible units. When machinery movement is necessary, the time of its occurrence can be manipulated to avoid periods of darkness or heavy vehicular traffic. By carefully scheduling the order in which tracts are to be farmed, miles can be shaved off the year's travel total. Duplicate sets of implements help cut down the amount of roading, too. Narrow bridges are modified (with and without authorization) or are bypassed with fords. Homemade trailers haul wide or unroadworthy implements. Thieves, rustlers, and vandals can be thwarted if appropriate steps are taken in advance.

Farm equipment manufacturers know the transport problems faced by their customers and have responded with a variety of implement modifications. To lift soil-working blades and the like off hard-surface roads, a special tractor hitch is available. Many big implements come equipped with their own pneumatic tires for road travel. Devices for reducing the width of equipment when on the road have been developed. Several models of implement trailer are on the market.

Tire makers have finally begun to attack the problem of excessive road wear on rear tractor tires. For years claims about tire durability were made, but traction bars continued to wipe off as farmers expanded their holdings and used hard roads to reach the outliers. In 1962, however, the two major tire firms, Goodyear and Firestone, announced revolutionary tread designs intended primarily to reduce scuffing.

From the public sector farmers have also received assistance. Local highway departments and other governmental officials at that level are widening narrow country bridges. State traffic codes are extremely lenient toward farmers and their implements of husbandry. Certain visible markings are required, but almost any size implement is permissible on midwestern roads. Furthermore, practically anyone may legally drive a tractor or combine on the road, even a preadolescent.

SELECTED BIBLIOGRAPHY

Local Documentation: Published Accounts

Cass County, Indiana: Official Rural Farm Directory, 1969. Algona, Iowa: Mio Directory Service Company, 1969.

Gentile, Richard J. Mineral Commodities of Putnam County. Report of Investigations 29. Rolla: Missouri Geological Survey and Water Resources, July, 1965.

Hays, John R. Relationship of Character of Farming Units to Land Management in Two Townships in Indiana. Bulletin 450. West Lafayette, Indiana: Purdue Agricultural Experiment Station, August, 1940.

Hunt, B. R.; Young, E. C.; and Robertson, Lynn. Land-Use Adjustments Needed on Farms in Deer Creek Township, Cass County, Indiana. Bulletin 466. West Lafayette, Indiana: Purdue Agricultural Experiment Station, February, 1942.

Illinois. Tax Commission. Atlas of Taxing Units, Vol. I of Survey of Local Finance in Illinois. Chicago: Illinois Tax Commission, 1939.

Joliet Army Ammunition Plant. Office of the Land Manager. Mailing List and Telephone Numbers of Agricultural Lessees. Elwood, Illinois, April, 1969. (Mimeographed.)

Joliet Herald-News. 1940-41 and other selected issues.

Kitchel, Mrs. Charles D. The History of the Daniel Crist Kitchel Family. Deer Creek Township, Cass County, Indiana: By the Author, n.d.

Kuzma, George J., ed. 100th Anniversary of St. Rose of Lima Parish: 1855-1955. Wilmington, Illinois: St. Rose of Lima Parish, [1955].

Logansport: City on the Grow. Logansport: Logansport Chamber of Commerce, [1969].

Logansport Pharos-Tribune and Press. 1938-70.

Peat, Marwick, and Livingston & Co. A Feasibility Study of Recreation and Tourism Development Potential of Lake Thunderhead and Putnam County, Missouri. N.p.: n.p., 1969.
This is a "technical assistance study" prepared under contract for the Economic Development Administration, U.S. Department of Commerce.

Peotone Vedette. 1938-69.

Robinson's 1964 Cass County, Indiana, Rural Directory. N.p.: Robinson Directories, Inc., 1964.

Unionville Republican. 1938-74.

U.S. Army. Corps of Engineers. Chicago District. Real Estate Division. *Agricultural Lease Program Summary*. Issued annually. 1949-64.

U.S. Department of Agriculture. Soil Conservation Service. *Soil Survey of Cass County, Indiana*, by L. R. Smith, W. J. Leighty, D. R. Kunkel, and A. T. Wiancko. Series 1939, No. 24. Washington: Government Printing Office, 1955.

Washer, H. L.; Veale, P. T.; and Odell, R. T. *Will County Soils*. Soil Report 80. Urbana: Illinois Agricultural Experiment Station, December, 1962.

Wiley, William G., comp. *Graduates from Young America High School, 1905-1963*. Flora, Indiana: By the Author, 1969. (Mimeographed.)

Local Documentation: Unpublished Sources

Agricultural Stabilization and Conservation Service offices in Joliet, Logansport, and Unionville. Current and historical files.

Carroll County, Indiana. *Deed Record*; *Discharge Record*; *Marriage Register*.

Cass County, Indiana. *Cemetery Deed Books*; *Deed Record*; *Discharge Record*; *Marriage Register*; *Miscellaneous Record*; *Tax Duplicates*.

Hays, John R. "Land Tenure and Land Use in Selected Areas in Indiana." Unpublished Ph.D. dissertation, Purdue University, 1940.

Howard County, Indiana. *Deed Record*; *Discharge Record*; *Marriage Register*.

Interviews with slightly more than 700 farmers, former farmers, friends and relatives of farmers, and local officials and businessmen during the period, February, 1969, to August, 1970.

Kankakee County, Illinois. *Deed Record*; *Tax Collector's Books*.

Manuscripts from the Illinois Tax Assessors' Annual Farm Census. State Archives. Springfield, Illinois. 1938-69.

Putnam County, Missouri. *Brand Record*; *Deed Record*; *Discharge Record*; *Marriage Register*.

Sullivan County, Missouri. *Deed Record*.

Tombstone inscriptions. Personal reconnaissance of numerous cemeteries in and around the study townships.

Tract indexes belonging to Ream Abstracts, Real Estate and Insurance (Unionville, Missouri): and Chicago Title and Trust (Joliet, Illinois).

U.S. Army. Corps of Engineers. Chicago District. Real Estate Division. Files pertaining to agricultural leasing at the Joliet Army Ammunition Plant. 1943-69.

Will County, Illinois. Deed Record; Tax Collector's Books.

Local Documentation: Cartographic Sources

Agricultural Stabilization and Conservation Service offices in Joliet, Logansport, and Unionville. Aerial photographs.

Carroll County, Indiana: Triennial Atlas and Plat Book. Rockford, Illinois: Rockford Map Publishers, Inc., 1968.

Cass County, Indiana. Surveyor's Office. Cass County, Indiana (Highway Map). 1967. Scale: 1:63, 360.

Cass-Howard Counties, Indiana: Plat Book and Index of Owners. LaPorte, Indiana: Town and Country Publishing Co., Inc., 1969.

Joliet Army Ammunition Plant. Office of the Land Manager. Land Utilization Maps. Issued irregularly. 1949-70.

Kankakee County, Illinois: Triennial Atlas and Plat Book. Rockford, Illinois: Rockford Map Publishers, Inc., 1967.

Kokomo, Howard County, Indiana. City-County Plan Commission. Road Numbering System--Howard County, Indiana. 1965. Scale: 1:63, 360.

Missouri. Highway Department. General Highway Maps.
Putnam County. 1964. Scale: 1:63, 360.
Sullivan County. 1965. Scale: 1:63, 360.
Putnam County. 1968. Scale: 1:126, 720.
Sullivan County. 1965. Scale: 1:126, 720.

Outdated plat books of Cass, Putnam, and Will counties. Found in various locations.

Putnam County, Missouri: Tri-Annual Plat Book. Marceline, Missouri: Advertising Enterprise, [1966].

U.S. Department of the Interior. Geological Survey. Topographic Quadrangles. Scale: 1:24, 000.
Anoka, Indiana (1955); Bunker Hill, Indiana (1962); Channahon, Illinois (1954); Clymers, Indiana (1962); Deer Creek, Indiana (1962); Elwood, Illinois (1953); Frankfort, Illinois (1946-53); Galveston, Indiana (1959); Lucerne, Missouri (1964); Manhattan, Illinois (1946-53); Miami, Indiana (1959); Onward, Indiana (1949-63); Pollock, Missouri (1964); Pollock NW, Missouri (1964); Pollock SW, Missouri (1964); Powersville, Missouri (1964); Symerton, Illinois (1953); Unionville West, Missouri (1964); Wilmington, Illinois (1954); Wilton Center, Illinois (1946-53); Young America, Indiana (1959).

Scale: 1:62,500.
Green City, Missouri (1912); Herscher, Illinois (1923); Kankakee, Illinois (1922); Peotone, Illinois (1946); Seymour, Iowa-Missouri (1942); Wilmington, Illinois (1954).

Scale: 1:250,000.
Aurora, Illinois (1958-68); Centerville, Iowa-Missouri (1954-67); Chicago, Illinois-Indiana-Michigan (1953-64); Danville, Illinois-Indiana (1953-65).

Will County, Illinois: Triennial Atlas and Plat Book. Rockford, Illinois: Rockford Map Publishers, Inc., 1969.

Items of a General Nature Pertaining to Agriculture

Advertisements issued in booklet or leaflet form by AMCO, American Products, Brillion, Deere, Donahue, Goodrich, Kewanee, Noble, Paul Abbott, Portable Elevator, and other firms producing equipment for farmers.

Aiken, Charles S. "The Fragmented Neoplantation: A New Type of Farm Operation in the Southeast." Southeastern Geographer 11 (April, 1971): 43-51.

Bailey, Charles H. "The Logistics of 35 Allotments." Farm Quarterly, Spring, 1970, pp. 64-65 and 90.

Birch, B. P. "Farmstead Settlement in the North American Corn Belt." Southampton Research Series in Geography 3 (November, 1966): 25-57.

Bremer, Richard G. "Patterns of Spatial Mobility: A Case Study of Nebraska Farmers, 1890-1970." Agricultural History 48 (October, 1974): 529-42.

Case, H. C. M., and Warren, S. I. "Why Tenants Move." Illinois Farm Economics 88 (September, 1942): 366-70.

Chester, Charles. "From 7 Farms to 1." Farm Quarterly, Winter, 1963-64, pp. 84-87.

Chisholm, Michael. Rural Settlement and Land Use. New York: John Wiley & Sons, Inc., 1967.

Clark, Neil M. "Then Came Rubber." Country Gentleman, November, 1939, pp. 14-15 and 70-71.

Clawson, Marion. Policy Directions for U.S. Agriculture: Long-Range Choices in Farming and Rural Living. Baltimore: Resources for the Future, 1968.

Clyde, A. W. "Disk Harrow Design Improvements." Agricultural Engineering 37 (March, 1956): 173-76.

Coppock, J. T. "The Geography of Agriculture." Journal of Agricultural Economics 19 (1968): 153-75.

Daily Pantagraph. Bloomington, Illinois. Selected issues.

Diller, Robert. Farm Ownership, Tenancy, and Land Use in a Nebraska Community. Chicago: The University of Chicago Press, 1941.

Dugan, Jack T. "A Geographic Analysis of Some Aspects of Land Tenure in Harlan County, Nebraska." Unpublished M. S. thesis, University of Nebraska, 1969.

Duncan, Otis Durant. "The Theory and Consequences of Mobility of Farm Population." Population Theory and Policy. Edited by Joseph J. Spengler and Otis Dudley Duncan. Glencoe, Illinois: The Free Press, 1956.

Edwards, Everett E. "Agricultural Records: Their Nature and Value for Research." Agricultural History 13 (January, 1939): 1-12.

Eisgruber, Ludwig M. "Changes in Farm Organization in a Central Indiana Township from 1910 to 1955." Unpublished M. S. thesis, Purdue University, 1957.

Farm Journal. 1939-69 plus selected earlier and later issues.

"Farms Unfrozen." Business Week, August 8, 1942, pp. 44 and 46.

Fisher, James S. "Federal Crop Allotment Programs and Responses by Individual Farm Operators." Southeastern Geographer 10 (November, 1970): 47-58.

Forrest, P. J. "Effects of Improper Inflation Pressures on Farm Tractor Tires." Agricultural Engineering 35 (December, 1954): 853-54.

Gasson, Ruth. The Influence of Urbanization on Farm Ownership and Practice. Studies in Rural Land Use, Report No. 7. Ashford, Kent: Department of Agricultural Economics, Wye College, 1966.

Graznak, Michael. "Violence Against Rural Property." Farmland, January 30, 1971, pp. 6-7.

Gregor, Howard F. Geography of Agriculture: Themes in Research. Englewood Cliffs, New Jersey: Prentice-Hall, 1970.

Haken, William ten. "Land Tenure in Walnut Grove Township, Knox County, Illinois." The Journal of Land and Public Utility Economics 4 (February, 1928): 13-24.

Haystead, Ladd. "Biggest Farmer Biggest No More." Fortune, December, 1943, p. 74.

Headington, R. C., and Falconer, J. I. Size of Farm Units as Affected by the Farming of Additional Land. Bulletin 637. Wooster: Ohio Agricultural Experiment Station, October, 1942.

Higgins, F. Hal. "Rural Revolution on Rubber." Pacific Rural Press, August 26, 1939; pp. 110-11.

Hughes, Robert C. "An Application of Markov Chains to Tenure and Size of Farm." Unpublished M. S. thesis, University of Illinois, 1962.

Hunt, Donnell. Farm Power and Machinery Management. Ames: Iowa State University Press, 1968.

"If You're Moving Next Spring." Wallace's Farmer and Iowa Homestead, October 18, 1952, p. 18.

Implement and Tractor. Selected issues.

Januszewski, Jozef. "Index of Land Consolidation as a Criterion of the Degree of Concentration." Geographia Polonica 14 (1968): 291-96.

Jensen, Clarence W., and Nash, Darrel A. Farm Unit Dispersal: A Managerial Technique to Reduce the Variability of Crop Yields. Technical Bulletin 575. Bozeman: Montana Agricultural Experiment Station, April, 1963.

Johannsen, Bruno B. "Tractor Hitches and Hydraulic Systems." Agricultural Engineering 35 (November, 1954): 789-93 and 800.

Kaups, Matti, and Mather, Cotton. "Eben: Thirty Years Later in a Finnish Community in the Upper Peninsula of Michigan." Economic Geography 44 (January, 1968): 57-70.

Kelsey, Myron P. "Economic Effects of Field Renting on Resource Use on Central Indiana Farms." Unpublished Ph. D. dissertation, Purdue University, 1959.

"Kewanee Disc Has Hydraulic Control." Farm Implement News, November 25, 1950, p. 38.

Kiefer, Wayne E. Rush County, Indiana: A Study in Rural Settlement Geography. Geographic Monograph 2. Bloomington, Indiana: Department of Geography, Indiana University, 1969.

Lively, C. E. "Spatial Mobility of the Rural Population with Respect to Local Areas." American Journal of Sociology 63 (July, 1937): 89-102.

McCuen, G. W., and Silver, E. A. Rubber-Tired Equipment for Farm Machinery. Bulletin 556. Wooster: Ohio Agricultural Experiment Station, October, 1935.

McGinnis, H. C. "Big Ones Eat the Little Ones in Machine-Made Farm Crisis." America, July 11, 1942, pp. 374-76.

McKain, Walter C., Jr., and Metzler, William H. "Measurement of Turnover and Retirement of Farm Owners and Operators." Rural Sociology 10 (March, 1945): 73-76.

"Magic Carpet." Saturday Evening Post, December 31, 1938, p. 22.

Malin, James C. The Grassland of North America: Prolegomena to its History. Lawrence, Kansas: By the Author, 1947.

________. "Mobility and History: Reflections on the Agricultural Policies of the United States in Relation to a Mechanized World." Agricultural History 17 (October, 1943): 177-91.

Malin, James C. "The Turnover of Farm Population in Kansas." The Kansas Historical Quarterly 4 (November, 1935): 339-72.

"Missouri Mud." Colliers, May 5, 1951, pp. 28-29.

Murray, William G. Farm Appraisal and Valuation. 5th ed. Ames: Iowa State University Press, 1969.

Nalson, J. S. Mobility of Farm Families: A Study of Occupational and Residential Mobility in an Upland Area of England. Manchester: Manchester University Press, 1968.

"New Tread Design for Tractor Tire." Wallace's Farmer and Iowa Homestead, June 2, 1962, p. 48.

Pike, Herbert. "How to Rent a Better Farm." Successful Farming, October, 1963, pp. 36 and 58.

Prunty, Merle, Jr. "The Census on Multiple-Units and Plantations in the South." The Professional Geographer 8 (September, 1956): 2-5.

Rikkinen, Kalevi. "Kalevala, Minnesota: Agricultural Geography in Transition." Acta Geographica (Helsinki) 19 (1969): 1-58.

Salter, Leonard A., Jr. A Critical Review of Research in Land Economics. Madison: University of Wisconsin Press, 1967.

________. Land Tenure in Process: A Study of Farm Ownership and Tenancy in a Lafayette County Township. Research Bulletin 146. Madison: Wisconsin Agricultural Experiment Station, February, 1943.

Schoeff, Robert W., and Robertson, Lynn S. Agricultural Changes from 1910 to 1945 in a Central Indiana Township. Bulletin 524. West Lafayette, Indiana: Purdue Agricultural Experiment Station, 1947.

Schwart, R. B. Farm Machinery Economic Decisions. Circular 1065. Urbana: Illinois Cooperative Extension Service, December, 1972.

"Shall I Move to Another Farm?" Wallace's Farmer and Iowa Homestead, August 1, 1953, p. 35.

Skovold, F. J. "Farm Loans and Farm Management by the Equitable Life Assurance Society of the United States." Agricultural History 30 (July, 1956): 114-19.

Smith, Everett G., Jr. "Road Functions in a Changing Rural Environment." Unpublished Ph. D. dissertation, University of Minnesota, 1962.

Tarver, James Donald. "Wisconsin Century Farm Families: A Study of Farm Succession." Unpublished Ph. D. dissertation, University of Wisconsin, 1950.

Taylor, Paul Schuster. "Good-By to the Homestead Farm: The Machines Advance in the Corn Belt." Harper's Magazine, May, 1941, pp. 589-97.

Timmons, John F. Improving Farm Rental Arrangements in Iowa. Research Bulletin 393. Ames: Iowa Agricultural Experiment Station, January, 1953.

"Two Centuries on this Farm." Hoard's Dairyman, June 10, 1942, p. 312.

U.S. Department of Agriculture. Power to Produce, The Yearbook of Agriculture, 1960. Washington: Government Printing Office, 1960.

________. Bureau of Agricultural Economics. Turn-over of Farm Owners and Operators, Vale and Owyhee Irrigation Projects, by Walter C. McKain, Jr. and H. Otto Dahlke. Berkeley, California: n.p., 1946.

________. Economic Research Service. Economies of Size in Farming, by J. P. Madden. Agricultural Economic Report 107. Washington: Government Printing Office, 1967.

________. Economic Research Service. Midwestern Corn Farms: Economic Status and the Potential for Large and Family-Sized Units, by Kenneth R. Krause and Leonard R. Kyle. Agricultural Economic Report 216. Washington: Government Printing Office, 1971.

U.S. Department of Transportation. Agricultural Tractor Safety on Public Roads and Farms. Washington: Government Printing Office, 1971.

Wakeley, Ray E. Differential Mobility Within the Rural Population in 18 Iowa Townships, 1928 to 1935. Research Bulletin 249. Ames: Iowa Agricultural Experiment Station, December, 1938.

Wardle, Norval J. Operating Farm Tractors and Machinery Safely and Efficiently. Pm-450. Ames: Iowa State University Cooperative Extension Service, March, 1969.

________. "Traffic + Tractors = Trouble." Iowa Farm Science 15 (May, 1961): 10-12.

Williams, W. M. A West Country Village, Ashworthy: Family, Kinship and Land. London: Routledge & Kegan Paul, 1963.

Wolf, C. E. "March 1st (What a Mere Date Can Mean to a Tenant Family)." Commonweal 31 (March 1, 1940): 404-5.

Other References

Adams, John S. "Directional Bias in Intra-Urban Migration." Economic Geography 45 (October, 1969): 302-23.

Boyce, Ronald R. "Residential Mobility and its Implications for Urban Spatial Change." Proceedings of the Association of American Geographers 1 (1969): 22-26.

Golant, Stephen M. The Residential Location and Spatial Behavior of the Elderly: A Canadian Example. Chicago: Department of Geography, The University of Chicago, 1972.

Junker, Buford H. *Field Work: An Introduction to the Social Sciences*. Chicago: The University of Chicago Press, 1960.

Lansing, John B.; Clifton, Charles Wade; and Morgan, James N. *New Homes and Poor People: A Study of Chains of Moves*. Ann Arbor, Michigan: Institute for Social Research, 1969.

Moore, Eric G. "The Nature of Intra-Urban Migration and Some Relevant Research Strategies." *Proceedings of the Association of American Geographers* 1 (1969): 113-16.

Packard, Vance. *A Nation of Strangers*. New York: David McKay Company, Inc., 1972.

Rossi, Peter H. *Why Families Move: A Study in the Social Psychology of Urban Residential Mobility*. Glencoe, Illinois: The Free Press, 1955.

Simmons, James W. "Changing Residence in the City: A Review of Intraurban Mobility." *Geographical Review* 58 (October, 1968): 622-51.

THE UNIVERSITY OF CHICAGO

DEPARTMENT OF GEOGRAPHY

RESEARCH PAPERS (Lithographed, 6×9 Inches)

(Available from Department of Geography, The University of Chicago, 5828 S. University Ave., Chicago, Illinois 60637. Price: $5.00 each; by series subscription, $4.00 each.)

84. KANSKY, K. J. *Structure of Transportation Networks: Relationships between Network Geometry and Regional Characteristics* 1963. 155 pp.
91. HILL, A. DAVID. *The Changing Landscape of a Mexican Municipio, Villa Las Rosas, Chiapas* NAS-NRC Foreign Field Research Program Report No. 26. 1964. 121 pp.
94. MC MANIS, DOUGLAS R. *The Initial Evaluation and Utilization of the Illinois Prairies, 1815–1840* 1964. 109 pp.
97. BOWDEN, LEONARD W. *Diffusion of the Decision to Irrigate: Simulation of the Spread of a New Resource Management Practice in the Colorado Northern High Plains* 1965. 146 pp.
98. KATES, ROBERT W. *Industrial Flood Losses: Damage Estimation in the Lehigh Valley* 1965. 76 pp.
102. AHMAD, QAZI. *Indian Cities: Characteristics and Correlates* 1965. 184 pp.
103. BARNUM, H. GARDINER. *Market Centers and Hinterlands in Baden-Württemberg* 1966. 172 pp.
105. SEWELL, W. R. DERRICK, et al. *Human Dimensions of Weather Modification* 1966. 423 pp.
106. SAARINEN, THOMAS F. *Perception of the Drought Hazard on the Great Plains* 1966. 183 pp.
107. SOLZMAN, DAVID M. *Waterway Industrial Sites: A Chicago Case Study* 1967. 138 pp.
108. KASPERSON, ROGER E. *The Dodecanese: Diversity and Unity in Island Politics* 1967. 184 pp.
109. LOWENTHAL, DAVID, et al. *Environmental Perception and Behavior.* 1967. 88 pp.
110. REED, WALLACE E. *Areal Interaction in India: Commodity Flows of the Bengal-Bihar Industrial Area* 1967. 210 pp.
112. BOURNE, LARRY S. *Private Redevelopment of the Central City: Spatial Processes of Structural Change in the City of Toronto* 1967. 199 pp.
113. BRUSH, JOHN E., and GAUTHIER, HOWARD L., JR. *Service Centers and Consumer Trips: Studies on the Philadelphia Metropolitan Fringe* 1968. 182 pp.
114. CLARKSON, JAMES D. *The Cultural Ecology of a Chinese Village: Cameron Highlands, Malaysia* 1968. 174 pp.
115. BURTON, IAN; KATES, ROBERT W.; and SNEAD, RODMAN E. *The Human Ecology of Coastal Flood Hazard in Megalopolis* 1968. 196 pp.
117. WONG, SHUE TUCK. *Perception of Choice and Factors Affecting Industrial Water Supply Decisions in Northeastern Illinois* 1968. 96 pp.
118. JOHNSON, DOUGLAS L. *The Nature of Nomadism* 1969. 200 pp.
119. DIENES, LESLIE. *Locational Factors and Locational Developments in the Soviet Chemical Industry* 1969. 285 pp.
120. MIHELIC, DUSAN. *The Political Element in the Port Geography of Trieste* 1969. 104 pp.
121. BAUMANN, DUANE. *The Recreational Use of Domestic Water Supply Reservoirs: Perception and Choice* 1969. 125 pp.
122. LIND, AULIS O. *Coastal Landforms of Cat Island, Bahamas: A Study of Holocene Accretionary Topography and Sea-Level Change* 1969. 156 pp.
123. WHITNEY, JOSEPH. *China: Area, Administration and Nation Building* 1970. 198 pp.
124. EARICKSON, ROBERT. *The Spatial Behavior of Hospital Patients: A Behavioral Approach to Spatial Interaction in Metropolitan Chicago* 1970. 198 pp.
125. DAY, JOHN C. *Managing the Lower Rio Grande: An Experience in International River Development* 1970. 277 pp.
126. MAC IVER, IAN. *Urban Water Supply Alternatives: Perception and Choice in the Grand Basin, Ontario* 1970. 178 pp.
127. GOHEEN, PETER G. *Victorian Toronto, 1850 to 1900: Pattern and Process of Growth* 1970. 278 pp.
128. GOOD, CHARLES M. *Rural Markets and Trade in East Africa* 1970. 252 pp.
129. MEYER, DAVID R. *Spatial Variation of Black Urban Households* 1970. 127 pp.
130. GLADFELTER, BRUCE. *Meseta and Campiña Landforms in Central Spain: A Geomorphology of the Alto Henares Basin* 1971. 204 pp.
131. NEILS, ELAINE M. *Reservation to City: Indian Urbanization and Federal Relocation* 1971. 200 pp.

132. MOLINE, NORMAN T. *Mobility and the Small Town, 1900–1930* 1971. 169 pp.

133. SCHWIND, PAUL J. *Migration and Regional Development in the United States, 1950–1960* 1971. 170 pp.

134. PYLE, GERALD F. *Heart Disease, Cancer and Stroke in Chicago: A Geographical Analysis with Facilities Plans for 1980* 1971. 292 pp.

135. JOHNSON, JAMES F. *Renovated Waste Water: An Alternative Source of Municipal Water Supply in the U.S.* 1971. 155 pp.

136. BUTZER, KARL W. *Recent History of an Ethiopian Delta: The Omo River and the Level of Lake Rudolf* 1971. 184 pp.

137. HARRIS, CHAUNCY D. *Annotated World List of Selected Current Geographical Serials in English, French, and German* 3rd edition 1971. 77 pp.

138. HARRIS, CHAUNCY D., and FELLMANN, JEROME D. *International List of Geographical Serials* 2nd edition 1971. 267 pp.

139. MC MANIS, DOUGLAS R. *European Impressions of the New England Coast, 1497–1620* 1972. 147 pp.

140. COHEN, YEHOSHUA S. *Diffusion of an Innovation in an Urban System: The Spread of Planned Regional Shopping Centers in the United States, 1949–1968* 1972. 136 pp.

141. MITCHELL, NORA. *The Indian Hill-Station: Kodaikanal* 1972. 199 pp.

142. PLATT, RUTHERFORD H. *The Open Space Decision Process: Spatial Allocation of Costs and Benefits* 1972. 189 pp.

143. GOLANT, STEPHEN M. *The Residential Location and Spatial Behavior of the Elderly: A Canadian Example* 1972. 226 pp.

144. PANNELL, CLIFTON W. *T'ai-chung, T'ai-wan: Structure and Function* 1973. 200 pp.

145. LANKFORD, PHILIP M. *Regional Incomes in the United States, 1929–1967: Level, Distribution, Stability, and Growth* 1972. 137 pp.

146. FREEMAN, DONALD B. *International Trade, Migration, and Capital Flows: A Quantitative Analysis of Spatial Economic Interaction* 1973. 202 pp.

147. MYERS, SARAH K. *Language Shift Among Migrants to Lima, Peru* 1973. 204 pp.

148. JOHNSON, DOUGLAS L. *Jabal al-Akhdar, Cyrenaica: An Historical Geography of Settlement and Livelihood* 1973. 240 pp.

149. YEUNG, YUE-MAN. *National Development Policy and Urban Transformation in Singapore: A Study of Public Housing and the Marketing System* 1973. 204 pp.

150. HALL, FRED L. *Location Criteria for High Schools: Student Transportation and Racial Integration* 1973. 156 pp.

151. ROSENBERG, TERRY J. *Residence, Employment, and Mobility of Puerto Ricans in New York City* 1974. 230 pp.

152. MIKESELL, MARVIN W., editor. *Geographers Abroad: Essays on the Problems and Prospects of Research in Foreign Areas* 1973. 296 pp.

153. OSBORN, JAMES. *Area, Development Policy, and the Middle City in Malaysia* 1974. 273 pp.

154. WACHT, WALTER F. *The Domestic Air Transportation Network of the United States* 1974. 98 pp.

155. BERRY, BRIAN J. L., et al. *Land Use, Urban Form and Environmental Quality* 1974. 464 pp.

156. MITCHELL, JAMES K. *Community Response to Coastal Erosion: Individual and Collective Adjustments to Hazard on the Atlantic Shore* 1974. 209 pp.

157. COOK, GILLIAN P. *Spatial Dynamics of Business Growth in the Witwatersrand* 1975. 143 pp.

158. STARR, JOHN T., JR. *The Evolution of Unit Train Operations in the United States: 1960–1969—A Decade of Experience* 1975.

159. PYLE, GERALD F. *The Spatial Dynamics of Crime* 1974. 220 pp.

160. MEYER, JUDITH W. *Diffusion of an American Montessori Education* 1975. 109 pp.

161. SCHMID, JAMES A. *Urban Vegetation: A Review and Chicago Case Study* 1975.

162. LAMB, RICHARD. *Metropolitan Impacts on Rural America* 1975.

163. FEDOR, THOMAS. *Patterns of Urban Growth in the Russian Empire during the Nineteenth Century* 1975.

164. HARRIS, CHAUNCY D. *Guide to Geographical Bibliographies and Reference Works in Russian or on the Soviet Union* 1975. 496 pp.

165. JONES, DONALD W. *Migration and Urban Unemployment in Dualistic Economic Development* 1975.

166. BEDNARZ, ROBERT S. *The Effect of Air Pollution on Property Value* 1975. 118 pp.

167. HANNEMANN, MANFRED. *The Diffusion of the Reformation in Southwestern Germany, 1518-1534* 1975.

168. SUBLETT, MICHAEL D. *Farmers on the Road. Interfarm Migration and the Farming of Noncontiguous Lands in Three Midwestern Townships, 1939-1969* 1975. 228 pp.